Innovation in Construction

Innovation in Construction

A European Analysis

Marcela Miozzo

Senior Lecturer in Innovation Studies, Manchester School of Management, University of Manchester Institute of Science and Technology (UMIST), UK

Paul Dewick

Lecturer in Technology Management, Manchester School of Management, University of Manchester Institute of Science and Technology (UMIST), UK

Edward Elgar
Cheltenham, UK • Northampton, MA, USA

Published by
Edward Elgar Publishing Limited
Glensanda House
Montpellier Parade
Cheltenham
Glos GL50 1UA
UK

Edward Elgar Publishing, Inc.
136 West Street
Suite 202
Northampton
Massachusetts 01060
USA

A catalogue record for this book
is available from the British Library

Library of Congress Cataloguing in Publication Data

Miozzo, Marcela, 1963-
    Innovation in construction : a European analysis / Marcela Miozzo, Paul Dewick.
        p. cm.
    Includes bibliographical references and index.
        1. Construction industry–Technological innovations–Europe. 2. Sustainable development–Europe. I. Dewick, Paul, 1975- II. Title.

HD9715.E82M56 2004
338.4'7624'094–dc22

2003068785

ISBN 1 84376 521 7

Printed and bound in Great Britain by MPG Books Ltd, Bodmin, Cornwall

Contents

Figures

Tables

Acknowledgements

This book arises out of a project funded by Scottish Homes, as part of EC Competitive Renewal Initiatives in Sustainable Europe Network Group. Our first acknowledgement is to the support of Scottish Homes for enabling this research to be conducted. Equal thanks are due of course to all the persons who offered us their time for interviews in the course of the research project in the five countries studied. Also, we would like to thank Alison Smith for her help in preparing the book. Finally, we would like to thank Dymphna Evans of Edward Elgar for her support.

Chapter 1 has been published in similar form as M. Miozzo and P. Dewick (2002), 'Building competitive advantage: innovation and corporate governance in European construction', *Research Policy*, **31** (6), 989–1008. Chapter 2 has been published in similar form as M. Miozzo and P. Dewick (2004), 'Networks and innovation in European construction: benefits from inter-organizational co-operation in a fragmented industry', *International Journal of Technology Management*, **27** (1), 68–92. We gratefully acknowledge permission from Inderscience to reproduce this material. Chapter 3 has been published in similar form as P. Dewick and M. Miozzo (2002), 'Sustainable technologies and the innovation–regulation paradox', *Futures*, **34** (9–10), 823–40. We gratefully acknowledge permission from Elsevier to reproduce the above material. Chapter 4 has been published in similar form as P. Dewick and M. Miozzo (2002), 'Factors enabling and inhibiting sustainable technologies in construction: the case of active solar heating systems', *International Journal of Innovation Management*, **6** (3), 257–72. We are grateful for permission from World Scientific Publishing Co to reproduce this material. Finally, Chapter 5 has been published in similar form as P. Dewick and M. Miozzo (2004), 'Networks and innovation: sustainable technologies in Scottish social housing', *R&D Management*, **34** (3), 323–33. We gratefully acknowledge permission from Blackwells to reproduce this material.

Introduction

Despite the economic significance of the construction industry and despite it being the creator of the built environment within which most other economic activity takes place, there are few scholarly analyses of innovation in construction and even fewer internationally comparative analyses. This may, in part, reflect the fact that construction has a poor public image in many countries. There is a popular perception that the diffusion of innovation is slow in the sector and that firms in the construction industry are excessively conservative and do not appreciate the benefits of technological innovation. In fact, there has been no absence of major technological developments in construction since the 1950s. These include new materials, prefabrication of modular components, industrialization of buildings, on-site mechanization, improved building services, application of EDI, CAD, CIM, and new structural solutions. These innovations, together with environmentally-responsible house-building and renovation, can bring important benefits to the wider economy. Nevertheless, the rate of adoption of innovation remains slow, and the view of the sector as conservative persists. There is a need, therefore, to understand the factors that enable and hinder innovation in the construction industry. Moreover, such investigation ought to be framed by guiding theories to interpret, organize and evaluate the pressures and challenges facing the sector.

This book presents the result of a project supported by Scottish Homes as part of the EC Competitive Renewal Initiatives in Sustainable Europe Network Group. We identify the key features of innovation in construction and the various opportunities and limitations to innovation in the sector, as affected by the nature of corporate governance, inter-firm relations and regulation. The book also explores the innovation process in relation to two specific technologies: natural thermal insulation and active solar heating systems. These technologies have been selected on the grounds that they may be expected to make a significant contribution to sustainable building and regeneration on their own account and that they have the potential to demonstrate at a more general level the underlying factors that facilitate or hinder the innovation process.

The results of this project are informed by almost seventy interviews with senior managers from contractors, housing associations and other clients, architects, engineers, professionals, representatives of government and quasi-government bodies, and professional institutions in five European countries: Denmark, France, Germany, Sweden and the UK. The interviews sought to compare the different ownership and governance structures of the main construction firms in each country and the networks between contractors, the government, suppliers of materials and machinery, and universities. The project analyses the structure of the sector by focusing on the forms of organization and co-ordination and their impact on its techno-economic performance.

The effect of the following factors in relation to innovation are evaluated in detail:

- ownership and organization structure of the leading construction firms;
- contractual forms and established methods of working;
- the nature of inter-firm co-operation, relation between construction firms and professionals and subcontracting linkages;
- the nature and quality of the interfaces and support that firms receive from government and other institutions at regional, local, national and international levels (in terms of infrastructure and support for collaboration);
- the impact of local and national standards and regulation; and
- the balance in final decision-making between short-term capital costs and long-term costs and benefits to the wider community.

We provide extensive empirical evidence and theoretical elaboration on two main points, which we argue are weaknesses of the 'systems of innovation' approach. These two points provide the main organizing theme for the whole book. These are as follows:

1. The Effect of Corporate Strategy and Structure on Innovation

Empirical research on innovation has neglected issues of corporate strategy and structure. Although the 'systems of innovation' literature includes the internal organization of the firm and financial institutions as factors that shape learning and innovation, there is little elaboration on how differences in patterns of ownership, finance and management and organizational structures contribute to the generation of process and product innovation. This book provides international comparative evidence on the effect of the structure of ownership and management on innovation in the construction industry. Also, it shows that particular structures of ownership and management, namely, concentration of ownership, cross-holdings and decentralization of management, which characterises the Germanic as opposed to the Anglo-Saxon corporate governance system, may generate the institutional conditions to support the commitment to irreversible investments in (firm-specific) innovation in construction firms.

2. The Importance of Inter-Organizational Networks for Innovation

Although the 'systems of innovation' literature acknowledges that the process of innovation is distributed between and within agents, it has a long tradition of treating the individual firm as an innovating entity. Innovation studies have paid inadequate attention to relationships between agents. The construction process may be regarded as an archetypal network system, since construction projects are planned and executed in the context of inter-organizational decisions, relations and activities. However, many of the problems of the performance of the construction industry stem from inadequate inter-organizational co-operation. We demonstrate this argument in this book by comparing the relations of contractors with subcontractors and suppliers, government, universities, clients and their international collaborations in the five European countries. In brief, in countries where inter-organizational relations are strong, such as in Denmark and Sweden, the productivity of the construction industry is higher, despite high labour and material costs.

The two points above not only help to advance a theoretical approach informed by the 'systems of innovation' literature but also enable an empirical assessment of the process of innovation in construction. As such, Chapters 1 and 2 explore the links between corporate governance and networks and the types of organizational strategies leading to innovation in processes and products in the five European countries' contractors. Chapters 3, 4 and 5 apply this framework to understanding 'systems of innovation' in the design and implementation of sustainable technologies. The factors that inhibit the adoption of technologies in the construction industry tend to be exaggerated when sustainable technologies are considered. To remedy this problem, targeted government policy is required, as well as improved inter-industry and intra-industry collaboration and wider public support. Even then, the empirical evidence shows that the difficulties in reconciling the different interests of the parties in inter-organizational networks are formidable. Despite the scope for greater sustainability in building, both in terms of sustainable processes (for example, waste minimization, recycling and sustainable design) and products (for example, high-tech renewable technologies and low-tech low-energy-embodied materials), the adoption of sustainable technologies varies across countries. This reflects cross-national differences in the type and extent of barriers to innovation. Overall, the book examines the role of different parties in the inter-organizational networks in facilitating and hindering sustainable innovation and the effect of government policy and other institutional initiatives to encourage the use of sustainable technologies.

This introduction sets out the background for the book. The first section explores the 'systems of innovation' literature and its relevance to the understanding of construction innovation. The second section examines the particular challenges posed to innovation studies by the study of sustainable technologies in construction. The third section examines the method adopted in the research project. The fourth section lays out the content of the rest of the book.

'SYSTEMS OF INNOVATION' AND THE CONSTRUCTION INDUSTRY

The 'systems of innovation' literature can provide the basis for a study of construction innovation. At the same time, however, the particular features of the construction sector pose challenges to innovation studies. In fact, despite the rise of the literature on 'systems of innovation', innovation studies has a long tradition of treating the individual firm as the innovating entity (Coombs et al. 2004). Little attention has been paid to the nature of inter-organizational relations, including the mechanisms through which economic co-ordination is achieved, competition is organized and regulated at different levels, and how rival arrangements compare and how this may influence the patterns of provision of goods and services and innovation. Also, little attention has been paid to the internal organization of the firm and how features of firm owner-ship, finance, organizational and management structures affect innovation performance.

The literature on innovation studies is still struggling to understand the linkages between innovation networks and the performance of the firms that participate in these networks. Also, within this broader problem, there is the specific issue of corporate governance and firm performance. The cross-national and longitudinal variability in the institutional forms of corporate governance are seen by some scholars as having significant bearing on firm financial performance; but the connections to innovative potential or performance are less well examined. These are under-researched elements in the dynamics of innovation networks or systems.

This is not to suggest that innovation studies have not provided important insights. Early work by SPRU (Science Policy Research Unit) (Rothwell 1977) gave strong support to the idea that success in innovation has to do with long-term relationships and close interaction with agents external to the firm. The work by von Hippel (1988) and Lundvall (1985 and 1988) highlighted the importance of user–supplier relationships in the innovation processes. These contributions, together with developments in evolutionary economics, provide a basis for the 'systems of innovation' perspective. The 'national systems of innovation' approach (Edquist 1997, Freeman 1987, Lundvall 1992, Nelson 1993) has underlined the network of institutions in the private and public sector whose activities and interaction initiate, import, modify and diffuse new technologies. Nevertheless, while this approach stresses the processes of interactive learning between institutions, including the production, marketing and finance system, in practice emphasis has tended to be placed on the scientific and knowledge infrastructure and there is little elaboration on how the different parties of the network interact with one another.

Another strand of innovation studies, the 'sectoral systems' literature (Breschi et al. 2000, Malerba 2002, Malerba and Orsenigo 1996) seeks to understand and compare the sources and patterns of technological change in different industries, focusing especially on learning processes and technological opportunities, appropriability conditions, cumulativeness in knowledge and the relevant knowledge base in an industry. However, the focus of these contributions tends to be the creation of new technology. Also, this literature may have problems in dealing with inter-sectoral interactions.

Insights are also derived from the literature on 'innovation in complex products and systems' (Davies and Brady 2000, Gann and Salter 2000, Hobday 2000, Prencipe 2000) which focuses on how innovation occurs in complex, high-value capital goods (such as telecommunication exchanges, aircraft engine control systems and intelligent buildings) produced by firms working together in projects. The scale and physical complexity of these products raise new innovation issues (previously unexplored for mass-produced goods). In particular, the features of these products lead to the inapplicability of conventional life-cycle models. They also point to the important role of tacit knowledge and other intangible assets. And, more importantly, because the span of managerial control may be outside the boundaries of the individual firm, collaboration is an important element of innovation in complex products and systems.

While many of these ideas can inform a study on construction innovation, the characteristics of the construction industry pose new challenges to the literature on 'systems of innovation'. The ability to develop new knowledge systematically and build on and renew scientific and technological competence has seemingly not been possible in construction (Gann 1994, 1997, Pries and Janszen 1995). Models of innovation developed to interpret manufacturing do not apply to construction (Slaughter 2000), since most of these models (with the exception of the literature surveyed above) generally assume that innovations are generated by an internal R&D organization that chooses among a set of promising research options (Nelson and Winter 1982) and that innovations can be exploited through large-scale mass production (Abernathy and Utterback 1978).

One of the problems with the construction industry is that construction facilities are large, complex and long-lasting, and they are created and built by a temporary alliance of disparate organizations concentrated temporarily on a single project and affected by standards, codes, tests and provisions for consumer protection, safety and environmental awareness (Arditi et al. 1997, Slaughter 1998, Tatum 1986, 1987, Laborde and Sanvido 1994, Rosenfeld 1994). A second issue is that different sectors of construction (those building sophisticated urban offices, bridge building or housing maintenance) use fundamentally distinct technologies, resources and skills. It has been suggested that the description of construction as an 'industry' is unsuitable (Groak 1994) and that more attention has to be given to the construction project (Allinson 1993, Winch 1998). Others argue that construction is better viewed as a process, rather than an industry, (Gann 1994, Tatum 1987) which includes designing, constructing, maintaining and adapting the built environment. All agree that these activities involve a multitude of organizations from a range of different industrial sectors, working together in temporary coalitions on project-specific tasks. A key conclusion, therefore, is that it is the project-based nature of these activities that is important when considering innovation, because this creates discontinuities in the development of knowledge and its transfer within and between firms and from one project to the next.

Indeed, innovation studies regard construction (housing) as 'supplier-dominated' (Pavitt 1984). The majority of R&D is carried out by materials and components producers who develop products aimed at improving the performance of buildings and structures (Quigley 1982, Pries and Janszen 1995). Very little R&D is carried out with the aim of improving construction processes. It is becoming increasingly recognized by industry and government that this adversely affects the performance in the use of technologies developed 'up-stream' of integration, assembly and installation work carried out by project-based construction organizations (Gann 1997). Moreover, some large materials and components producers may be the source of major innovations for construction, but they may not regard construction as their primary market in terms of the focus on R&D efforts (for example, chemicals and glazing products). Firms with technical capabilities (of which there are only a handful in the construction sector) must 'intercept' or 'tap' technologies developed elsewhere in other industries or other countries and reconfigure them for specific purposes within their projects (Gann 1997, Toole 1998).

The above suggests that although insights can be gained from the 'systems of innovation' literature, particular features of the construction sector present new challenges for innovation studies. In particular, the project-based nature of work creates discontinuities in the development of knowledge within the firm and its transfer between firms and projects, suggesting the need to study in depth the effects on innovation of the internal organization of firms and their inter-organizational relations.

SUSTAINABLE INNOVATION IN THE CONSTRUCTION INDUSTRY

The problems in applying the lessons of innovation studies are even more acute when considering sustainable technologies. General studies in sustainable technologies draw insights that have some applicability for the construction industry. First, although technological innovation remains one of the most important factors in contributing to sustainability (Gray 1989, Green et al. 2002, IPCC 2000, Weaver et al. 2000, Weitzmann 1997), the development of technology on its own is unlikely to achieve a more sustainable future. The successful development and adoption of sustainable technologies requires regulation, economic incentives, private support and, in particular, an active role of the government (Freeman 1996, Green et al. 1994, Kemp 1994). Second, as with any new technology, the adoption of sustainable technologies is hindered by prevailing technological trajectories. In the case of sustainable (or 'cleaner') technologies, the dominance of prevailing technologies is even stronger (Kemp and Soete 1992).

There are yet further challenges posed by sustainable technologies in the construction industry. Sustainable innovation in the construction industry can be defined as changes associated with reducing the energy requirements of buildings and/or reducing the environmental impact (the so-called 'environmental footprint') of buildings and structures. This includes product innovation (for instance, in the use of natural materials, recycled/renewable materials or low embodied energy materials) and process innovation (for instance, resource-efficient construction methods such as the minimization of energy and waste, maximization of recycling, local sourcing of materials and the use of brownfield sites). But, to be truly sustainable, technologies must also have a social and economic dimension. The social dimension can be in terms of intra-generational equity, improving the standard of living of the poorer sectors of society (for example, by reducing the energy bills of social housing tenants) and the economic dimension can be in terms of not compromising the need for private firms to maintain certain levels of profit, particularly in a low-profit margin industry like construction.

A fourth problem concerning the diffusion of sustainable technologies in construction relates to the network of firms collaborating on a building project. The need to engage the entire network is particularly pertinent for the adoption of sustainable technologies since the client that specifies (and funds) the sustainable products and processes neither designs, manufactures, implements nor, in some cases, even uses the technology. This interdependency required for the effective diffusion of sustainable technologies and use of sustainable processes has been hindered in the past by a 'vicious cycle of blame', whereby each actor in the industry blames each other for not building environmentally friendly buildings (Cadman 1999).

Also, the construction industry is heavily regulated, including technical regulations, governing products and processes; planning and environmental regulations, governing the finished product; and health and safety regulations, governing the welfare of workers during the construction process (Gann 1999). Although some would claim that the extent and range of these regulations impose considerable restraints on technological innovation (Tatum 1987), there is evidence that, properly designed, regulations can act as a spur to sustainable development and the introduction of sustainable technologies (Kemp and Soete 1992, Porter and Van de Linde 1995).

Perhaps the main barrier relates to the perception of sustainable technologies as being inherently more risky than other technologies. In the construction industry this is predominantly a reflection of the costly and problematic nature of realizing an innovative specification. As we will see from our research, the higher costs stem from the additional development costs of the technology, the absence of economies of scale in production, the absence of economies of learning and experience in the implementation of a new technology, the lower number of manufacturers and the higher transport costs. Sourcing the technology also tends to be more problematic because of the difficulty in identifying designers, suppliers and sub-contractors with the capability, experience and willingness to design, supply and install the new technologies.

Finally, the government can be very influential in facilitating the achievement of sustainable targets through its role as largest single client of the building industry and by using fiscal and regulatory measures to stimulate innovation and act as a broker in markets for environmental technologies. Particularly with sustainable technologies, the government also has an important role as chief educator and disseminator of information (both to the industry and to the general public) and as market leader, prototyping innovative solutions through demonstration projects.

Overall, therefore, the challenge for the innovation literature is to acknowledge that the innovation process for sustainable technologies in construction is inherently complex since there are multi-dimensional considerations associated with their adoption. Economic incentives are important but consideration must also be applied to social and environmental aspects. Also, technological innovation is only one of the factors in contributing to sustainability; a similar degree of attention needs to be focused on organizational and institutional innovation. For sustainable technologies, the role for active government policy and information dissemination (from government and industry) cannot be overstated. This book addresses directly these concerns and offers an integrated approach to the assessment of sustainable technologies, with applicability to sectors other than the construction industry.

METHOD

Our discussions are informed by extensive interviews, especially focusing on the largest three to four contractors in each of the five countries and with a number of professional institutions, representatives of government, quasi-government bodies, research institutes, architects and clients in the five countries. We have thus left out the majority of firms that are small and medium-sized firms and those in design and engineering consultancy and supply industries. However, large contractors in the five countries exhibit wide differences in ownership and management structures, profitability, and forms of long-term relations. In line with our two main points providing the main organizing theme, we argue in the book that the focus on large contractors may be justified for three reasons:

- Contractors play an integrating role in the project and a mediator role in the interface between the institutions that develop many of the new products and processes (materials and components suppliers, specialist consultants and trade contractors) and those that adopt these innovations (clients, regulators and professional institutions) (Winch 1998). Unless the contractor can 'intercept' innovations developed elsewhere, such as new materials or process and has the skill to learn and apply it in future projects, as well as to incorporate it into the system as a whole, change is likely to be slow.
- Contractors are not only mediators in the project coalition but there is evidence that they are an important source of innovation to improve construction technology, and, more importantly, to integrate the different systems (Slaughter 1993). Evidence of this, however, is not universal. In several countries (most notably in the UK), it appears that corporate governance structures and the practice of awarding contracts through lowest-cost tender may act as a constraint to innovation among contractors. Under these circumstances, contractors may be left with little autonomy to alter design specifications and introduce product and process innovations.
- Contractors are also in a unique position for driving forward a sustainable path for the construction industry). Particularly in light of

recent developments (for example, the changing procurement forms and public sector tender requirements), the role of the contractor has become more important in facilitating the use of sustainable products and implementing sustainable processes. Large contractors face different challenges to small contractors, but, regardless of firm size, they must look beyond the costs and accept the liability and risk associated with building with sustainable technologies.

THE STRUCTURE OF THIS BOOK

The two chapters in Part I contribute to the literature of 'systems of innovation' through an examination of the relation between corporate structure and networks and innovation in the construction industry across five European countries. Chapter 1 explores the relationship between corporate governance and innovation in European construction. The ability to undertake research and development in production technologies by contractors differs widely across different countries. This may be explained by the extent to which strategic control is in the hands of those who have the incentives and abilities to allocate resources to uncertain and irreversible investments in innovation. This is influenced by particular features of firm ownership, organizational and management structure, internal mechanisms to diffuse knowledge within the firm and links to external sources of knowledge. Our evidence demonstrates that contractors with a Germanic corporate governance structure are in a better position to develop a long-term strategy of innovation. With other corporate governance systems, in the absence of the influence of a range of stakeholders (banks, industrial firms and workers) contractors are likely to be pressed into meeting the short-term interests of shareholders, rather than engage in long-term investment in production technology and machinery.

Chapter 2 examines the relation between inter-organizational networks and innovation in the construction industry. The performance of the construction industry differs widely across different countries. Our research findings suggest that the strength of inter-organizational co-operation may be responsible for enhanced performance of the construction industry in some of the countries. We examine the strength of the relationships of contractors with subcontractors, suppliers of materials, the government, universities, designers (architects or engineers), clients, and international collaborations with other contractors. In countries where inter-organizational relations are strong, such as Denmark and Sweden, the productivity of the construction industry is higher, despite high labour and material costs. In Denmark, the government has taken an active role in promoting collaborations along the building supply stream through demonstration projects to encourage process innovations. Also, architects and engineers are actively integrated into the supply stream and have an important role in designing and incorporating new products. In Sweden, longer-term relations between construction firms and universities and with material suppliers and manufactures are responsible for process and product innovation.

Part II examines the factors facilitating and hindering the diffusion of sustainable technologies in construction. Chapter 3 examines the paradox between innovation and regulation and the implications for the adoption of sustainable technologies in the domestic sector of the construction industry. The case of the UK is examined, where progress towards the inclusion of social and environmental considerations has been slow. Recent changes in contractors' concerns with environmental performance, combined with government initiatives, have prompted a more sustainable agenda in construction. With significant reductions in greenhouse gas emissions required to meet climate change targets, the case for a sustainable technology designed to reduce the energy required for space heating – natural thermal insulation materials for cavity wall insulation – suitable for widespread use in residential buildings, is assessed. Despite having lower embodied energy, natural thermal insulation materials do not match the energy-saving performance of the higher embodied energy incumbents. Energy savings from the use of thermal insulation far exceeds the energy savings associated with its production. This means that the incumbent technology is more 'sustainable'. Tighter building regulation is required to increase the minimum insulation levels and improve the sustainability of the housing stock.

Chapter 4 examines the factors enabling and inhibiting sustainable technologies in construction across Europe through a detailed analysis of the case of active solar heating systems. Active solar heating is a sustainable technology suitable for widespread use across new and existing buildings in the housing stock, which has the potential to make a significant contribution to sustainable building and regeneration. The generally slow adoption of this sustainable technology can be attributed to high capital cost and unknown cost effectiveness, but these factors do not adequately explain variations in its adoption across Europe. Indeed, considerable differences between European countries in the take-up of new sustainable technologies in construction suggests that there are sets of more important factors and institutions inhibiting or facilitating their adoption. This chapter examines the structural and institutional factors behind these differentials and draws implications for the management of innovation by construction firms and government policy for those countries under-exploiting the potential of active solar heating systems.

Chapter 5 examines inter-firm relations and sustainable technologies in the

Scottish social housing sector. The process of innovation often involves the participation of several firms and organizations. This chapter is concerned with improving our understanding of this process of innovation and inter-organizational relations by assessing the case of the introduction and diffusion of sustainable technologies in the Scottish social housing sector. Despite policy initiatives by the national housing agency to encourage sustainable technologies and processes, a number of factors related to the organizational form of the construction industry militate against the achievement of this objective. The different aims of the parties involved in the construction network may not be easily reconciled and traditional approaches to construction may reinforce these differences hindering efforts to introduce innovation.

PART I

Systems of innovation and the European
construction industry

1. Corporate governance and innovation in construction in five European countries

INTRODUCTION

It has been argued that different institutional frameworks have comparative advantages in solving the organizational problems of different innovation strategies (CPB 1997, OECD 1995). The general conclusion is that technological development through radical innovations may be encouraged by more market-oriented or Anglo-Saxon models of corporate governance while, in contrast, incremental technological change may be supported by network or Germanic models of corporate governance. This is because radical innovations make use of marketable assets, such as general human capital or external know-how, rather than firm-specific assets and knowledge that need to be developed internally, and demand flexible financial institutions and a high amount of risk finance (CPB 1997). Instead, incremental innovations shift the balance towards long-term finance opportunities to meet idiosyncratic customer requirements. Because banks, workers, governments and large shareholders have better information and more power to use that information than the widely dispersed shareholders of the typical UK or US firm, it is argued that financing for innovation is more readily available for value-increasing, long-term projects in the Germanic model. Other institutions such as vocational training reinforce the impact of these features of the Germanic model.

Missing from this analysis is an explanation of the particular relationships between corporate governance and the different types of innovative activities at the firm level. By examining the mechanisms of innovation at the firm level we are able to understand that although corporate governance systems may be broadly similar between groups of countries, differences in particular features of firm ownership, finance, organizational and management structures and mechanisms to diffuse knowledge within the firm may result in different types of innovation activities.

This chapter addresses these issues through a study of the construction industry in five European countries: Denmark, France, Germany, Sweden and

15

the UK. The project-based nature of work in the construction sector implies that firms have to manage networks of highly complex innovation interfaces. As such, construction can be viewed as a complex industry in which there are many interconnected and customized elements organized in a hierarchical way, with small changes to one element of the system leading to large changes elsewhere (Ball 1988, Gann 1994, Slaughter 1998). In this context, large contractors play a mediator role in the interface between the institutions that develop many of the new products and processes (materials and components suppliers, specialist consultants and trade contractors) and those which adopt these innovations (clients, regulators and professional institutions) (Winch 1998). Unless the contractor as mediator is convinced of the merit of the new material or process and has the skill to learn and apply it in future projects, as well as to incorporate it into the system as a whole, change is likely to be slow.

Contractors are not only mediators in the project coalition but, especially large contractors, can also be an important source of innovation to a much greater extent than is usually recognized (Slaughter 1993). For example, a recent survey in Germany found that approximately 60 per cent of contractors with 200 or more employees were innovative (developing either product or process innovations) (Cleff and Cleff 1999). Evidence of this relationship, however, is not universal. In several countries (most notably in the UK), it appears that the practice of awarding contracts through lowest-cost tender may act as a constraint to innovation and R&D spending among contractors (Ball 1996). Under this particular procurement form, contractors are left with little autonomy to alter design specifications and introduce product and process innovations.

This chapter explores the development of strategic innovations and operational capabilities in the largest contractors, measured by turnover, in each of the five countries. At each contractor, we conducted interviews with senior personnel and collected documentary evidence regarding research and development activities and information associated with particular features of corporate governance (see Table A.1 in the Appendix for details of firms interviewed). The empirical evidence from our 17 case-study contractors suggests that particular features of the corporate governance structure of contractors support different types of innovation at the firm level. For that purpose, the first section argues that research on the relationship between corporate governance and innovation has been limited. The second section explores the particular nature of this relationship in the construction industry. The third section describes our case study findings with respect to corporate governance and the innovative activities undertaken by the contractors. A final section discusses the research results and policy implications of this analysis.

CORPORATE GOVERNANCE AND INNOVATION

Research on the relationship between corporate governance and the process of innovation has been limited to date because the main theories of corporate governance do not integrate systematically an analysis of the economics of innovation. Corporate governance deals with the ways in which suppliers of finance to corporations act to ensure they achieve a return on their investment (Shleifer and Vishny 1997). The principal–agent framework plays a central role in several analyses of management incentives and accountability (stimulated by Berle and Means' (1932) concept of separation of ownership and control in the modern business enterprise) (Coase 1937, Fama and Jensen 1983, Jensen and Meckling 1976). The corporate governance literature provides illustrations of the variety of mechanisms to solve agency problems, including profit sharing, direct monitoring by boards, competition among managers, the capital market, and the market for takeovers. As such, it refers to the difficulties financiers have in assuring that their funds are not wasted on unattractive projects and considers the market and administrative checks designed to avoid this. However, it provides no systematic explanation of the conditions under which managers will make investments that promote or discourage innovation.

Indeed, as argued by O'Sullivan (2000a, 2000b), Anglo-American debates on corporate governance have been dominated by a shareholder theory, the main challenger to which is a stakeholder theory. Despite other differences, both these theories share the assumption of resource allocation as optimal and a focus on which party should lay claim to the residual if economic performance is to be enhanced. Little or no effort has been devoted to understanding how these residuals are generated through the development and utilization of productive resources.

In a similar fashion, most of the empirical research on innovation has ignored issues of corporate strategy and structure. Although the national systems of innovation literature include the internal organization of firms and financial institutions as factors which shape learning and innovation (Freeman 1987, Lundvall 1992, Nelson 1993), there is little elaboration on how differences in patterns of ownership, finance, and management and organizational structures contribute to the generation of process and product innovations. There are some notable exceptions, including contributions that have explored the effects of corporate governance on technological innovation and how variations in national systems of corporate governance can help explain national patterns of sectoral specialization (Lazonick and O'Sullivan 1996, Tylecote and Conesa 1999).

There is a need to bring together these two broad areas of study. The characteristics of innovation – localization, cumulativeness, firm-specificity and appropriability (Arthur 1988, Atkinson and Stiglitz 1969, David 1985, Nelson and Winter 1977, Teece 1986) – imply that innovation requires a sustained effort, the outcome of which is uncertain. A theory of corporate governance must therefore come to terms with the nature of innovation. It must explain how particular structures of ownership and management of firms generate the institutional conditions to support the commitment of resources to irreversible investments in innovation. The recent work of O'Sullivan (2000a, 2000b) provides a useful frame of reference in exploring this issue; and we use this as a starting point for our analysis. O'Sullivan conceptualizes innovation as a strategic issue. Similarly, questions of corporate governance are not so much a matter of whether profits go to shareholders or whether the interests of stakeholders are well represented; rather, for O'Sullivan, the focus is on the way differences in governance structures of firms shape the extent to which strategic control is in the hands of those with incentives and abilities to allocate resources to uncertain and irreversible investments in innovation. It is this concern with the way in which the interrelationship between corporate governance and innovation drives the development and utilization of productive resources that is at stake in this chapter. In the next section we set out the peculiarities of this relationship with regard to the construction sector.

CORPORATE GOVERNANCE AND INNOVATION IN CONSTRUCTION

This section outlines the particular features of innovation in construction and the way this is enabled and hindered by different forms of corporate governance. Traditionally, suppliers of materials and machinery were viewed as the main sources of innovation in construction (Pries and Janszen 1995, Quigley 1982). It has been argued that:

> the construction sector can be characterized by the great number of small enterprises and varying collaborations; co-makership (or other strategic alliances) hardly exists. The emphasis lies on operational (project) management. Strategic management does not exist ... commonly the horizon of contractors is not beyond the moment of completion of a project. ... (Pries and Janszen 1995, p.44)

However, other experts on innovation in construction have pointed out that general and speciality contractors are important sources of innovation, particularly for innovations that involve the integration and interaction among systems (Slaughter 1993). Also, it has been pointed out that there may be strong strategic company-wide incentives for and benefits from innovation even if the expected project-based benefits do not appear to offset the expected costs (Winch 1998).

Innovations can provide the critical component of a firm's competitive strategy. In this chapter we focus on the products (materials and components), processes and systems, specifically associated with the design and construction of built facilities (Slaughter 2000). The discontinuous and temporary nature of project-based modes of production in construction, however, may present a problem for the accumulation of knowledge. In construction, therefore, some of the most important issues include: the extent to which firms integrate the experience of projects into their business processes to ensure the coherence of the organization; the presence of institutions to capture knowledge and learn from past projects; the presence of a coherent technical support system at the core of the firm to support projects; and the mechanisms to capture knowledge from outside the firm (Gann and Salter 2000).

Our case study material confirms these observations and shows that contractors may be involved in two modes of innovative activities: research and development at a strategic level; and the development of operational capabilities. The first mode concerns research and development into products and processes that have a significant influence on the firm's future organization, development and strategy. Strategic research and development activities may stem from any level within the organization (top management, middle management or project management), or from external sources, but decisions related to its adoption are taken at the top level and involve large sums of funding. The development of operational capabilities can be divided into two types, the benefits of which are maximized through the creation of institutions within the organization to facilitate economies of experience and learning. The first concerns project-based innovations that stem from experience on site or incremental changes to existing processes or products. The second relates to the generic build-up of knowledge within the organization, generated through internal organization and in-house diffusion mechanisms and alliances and links to external sources of information and knowledge.[1]

Investments in process and product innovations are essential, but high costs and minimum efficient scales may make it difficult for firms to undertake R&D on their own. Differences in forms of corporate governance play an important role here. The particular structures of ownership and management are an important factor determining the ability to invest in innovation. In this chapter, we focus on the degree of concentration of ownership, the existence of cross-holdings and the extent to which the management structure is decentralized. The corporate governance structure of the contractor is particularly important for large-scale projects that require significant capital investment. For example, contractors may be more likely to invest in firm-specific assets or complementary knowledge if they can finance the project internally from cash flows or reserves. The capacity to retain earnings and channel profits toward investments (as opposed to toward dividends) will be significantly influenced by the corporate governance system. Less profitable firms, without the necessary internal capital to fund projects may need to borrow to finance capital investment and issue debt. Thus, corporate governance influences both the degree to which contractors are able to channel profits toward residual cash flow or research and development spending and the leverage they have to fund externally through debt issues. Indeed, where one of the principal owners is also the lender of last resort, such as a bank, firms may be able to access debt more easily for investment in firm-specific assets. A third way may be available to contractors to finance innovation, through sharing the costs of R&D in collaborations with other firms or participating in research programmes organized by national or European governments.

In some countries, contractors have relatively little fixed capital since, despite ownership of buildings and land, they do not own significant assets that could be used as collateral to access cheaper loans or that could be sold in the event of financial distress. Also, in some countries, contractors do not own factories manufacturing prefabricated components or active mining operations and much of the plant and machinery is outsourced. In these cases, the principal assets of a contractor are of an intangible nature; for example, a construction process know-how and an ability to manage various elements of the system efficiently. In addition to the intangible and inherently risky nature of contractors' assets, the 'safety' of their assets is also affected by cyclical movements in the economy. Moreover, the market value of specific firm divisions are determined by the strength of local markets in which they operate.[2] Hence, the organizational structure of the contractor (in terms of its internal organization into different functional divisions and degree of geographic diversification) also impinges upon its risk exposure and leverage capacity. Moreover, because of the high risks associated with potential outcomes of research and development and with the uses of innovation, the government can play an important role in guaranteeing public markets for innovating firms (Groenewegen 1994).

Corporate governance and the internal and external organizational structure of contractors also affect the development of operational capabilities. The firm's internal organizational structure (such as the level of decentralization and the mechanisms established to diffuse innovative ideas and best practice and to transfer knowledge throughout the organization) and external organizational structure (linkages with external sources of information and knowledge) play an important role in promoting incremental innovations. Indeed, incremental innovation is predominant in the construction industry (Gann 1994), characterized by an interactive process in which the main organizations involved provide and exchange different kinds of resources and goods (financial, human, information and material). Moreover, innovations in construction are not implemented within the firm itself but as part of the projects in which firms are engaged. Since these projects are collaborative engagements with other firms, most innovations have to be negotiated with one or more parties within the project coalition. In this sense therefore, incremental technological change may be supported by governance systems with cross-holdings among industrial firms, which may facilitate long-term relations between them.

CORPORATE GOVERNANCE AND INNOVATION OF CONTRACTORS IN FIVE EUROPEAN COUNTRIES

This section draws on case-study material to explore the relationship between corporate governance and innovation in five European countries. In each country, we identified the leading contractors in the construction industry and negotiated access for carrying out interviews with senior managers and project managers and the collection of documentary information. This resulted in the selection of the top four contractors in Denmark and Sweden and the top three in Germany, France and the UK (see Table A.1 in the Appendix).

The authors interviewed the directors of technology or heads of research and development in the contractors. Where specific research and development projects were in operation, project managers were interviewed to provide more detail. The interviews were conducted in a semi-structured form with core questions asked to each representative of the main contractors. Table A.2 in the Appendix shows that the leading contractors constitute a significant proportion of the total national construction industry's turnover and employment, differing in magnitude between Sweden and France, on one hand, and Denmark, Germany and the UK, on the other.

The following analysis identifies features of corporate governance in the top contractors, focusing on details of ownership structure, source of finance, degree of decentralization of management structure and the types of cross-holdings within the industry (see Table 1.1 for a summary). For each country, we assessed the impact of these features of corporate governance on innovation. As outlined above, innovation in construction includes both research and development at a strategic level and the development of operational capabilities. Because of the particular nature of the construction industry, it tends to involve collaborative relations with other firms. Comparative analysis of the research findings is presented in the final section.

Table 1.1 Effects of main features of corporate governance on innovation in large European contractors

Corporate governance features	Germany	Sweden	Denmark	France	UK
Ownership and control	Concentrated; combined influence of industrial firms, banks and workers facilitates long-term firm-specific investments in innovation	Combined influence of industrial firms, banks, family and workers facilitates long-term firm-specific investments in innovation	Smaller and wholly-owned subsidiaries; combined influence of family, foundations and workers facilitates long-term -firm specific investments in innovation	Concentrated family ownership facilitates long-term firm-specific investments in innovation	Institutional investors; concern with dividends leads to investments in project management and organizational innovation
Income derived from abroad	Relatively high stream of finance from abroad allows more stable long-term funding for innovation	High stream of finance from abroad allows more stable long-term funding for innovation	Low	High stream of income from abroad (but also emphasis on domestic market) allows more stable long-term funding for innovation	Relatively low (but increasing)

Management structure	Relatively decentralized, but central co-ordination of innovation	Decentralized, but very strong central co-ordination of innovation	Centralised management structure, control by parent firms	Decentralized, but strong central co-ordination of innovation	Relatively decentralized, but strong central co-ordination of innovation
Form of cross-holdings	Strong pattern of cross-holdings where suppliers and clients support collaboration for innovation	Strong pattern of cross-holdings where suppliers and clients support collaboration for innovation	Few; collaboration with government and project team	Strong pattern of cross-holdings where suppliers and competitors support collaboration for innovation	None

23

Germany

The leading German contractors interviewed (Holzmann, Hochtief and Strabag) are characterized by a well-developed system of cross-holdings with industrial firms and banks, and by a relatively concentrated structure of ownership of shares by both banks and non-financial firms (see Table 1.2).[3] Each contractor has a two-tier board system, in line with German legislation that makes it obligatory for large firms (over 2000 workers) to have employee representation on the board, with the supervisory board combining shareholder control with employee co-determination. Representatives of banks have to take care of the interests of the bank as shareholder, of the private shareholders that the bank may represent as proxy holder, and frequently also take into account that the bank may have a lending relationship with the firm. Representatives from non-financial firms may combine their interests as a block shareholder with supervision of a supplier relationship. As other studies have found, this combination of interests of many stakeholders in one institutional body may complicate decision-making but may also ensure that the risks and expected returns from long-term, and firm-specific innovation are better assessed and more readily financed (CPB 1997).

At Holzmann, Hochtief and Strabag, the combined interests and organizational integration of banks, non-financial firms and workers tends to support investment in firm-specific innovations that demand significant funding. For example, Hochtief has developed an integrated voice and data communication system for large construction projects, which includes a compact unit (the 'communications container') that integrates all mobile communications and IT components, linking the site with the firm's switchboard, servers, faxes and computers, connected to an external power supply and the ISDN network. At Holzmann, firm-specific strategic long-term projects, instead, have emphasized new building materials and machinery. Materials developments include high-strength concrete for the construction of high-rise buildings, towers or offshore structures and SIMCON, a layer of concrete reinforced with thin mats of steel for heavy-duty construction. Holzmann has also developed a non-destructive radar method of locating damage to concrete building material and a method of risk analysis which optimizes the use of shield boring machines to reduce technical risks in tunnelling. Despite the fact that two of the three German contractors experienced poor financial performance (neither Holzmann nor Strabag have paid dividends since the mid-1990s), the organizational integration and combined interests of stakeholders have ensured investment in long-term and firm-specific projects.

While the high degree of decentralization of management at the three contractors may be expected to hinder the diffusion of innovation, evidence of

Table 1.2 German contractors: ownership, control and structure

Company	Leading shareholders (% of shares)	Controlling interest	Income from abroad (%)	Active European holdings	Active overseas holdings
Philipp Holzmann	Foreign institutional investor (30); bank (21); dispersed ownership (34)	Foreign institutional investor; bank; domestic institutional investor	45	12	4
Hochtief	Domestic industrial firm (66); dispersed ownership (34)	Domestic industrial parent firm	47	5	5
Strabag	Foreign industrial firm (>50); dispersed ownership (<50)	Foreign industrial parent firm	41	22	0

Notes:
1. Active holdings represent more than 50 per cent share ownership of a second firm in another European or overseas country.
2. There are different opinions on the extent of ownership required to guarantee control. In theory, a single shareholder with 49 per cent of the capital could be outvoted by the other shareholders grouping together and block voting. However, in practice the figure is considerably less. Radice (1971) and Steer and Cable (1978) argued that a single shareholder owning 15 per cent would guarantee control. These 'cut-off' figures are too simplistic however, and one must look more closely at the distribution of ownership, considering for example: the identity of the largest shareholder, the extent of cross ownership and inter-locking directorships, the board of directors' share of equity, the number of family/founder members on the board, links with financial institutions and so forth (Nyman and Silberston, 1978).

Source: Individual firms' annual reports (1998), European International Contractors (1998).

strong coordination among divisions enables the dissemination of innovations arising from different projects. Indeed, Holzmann and Strabag have decentralized management structures across geographical divisions,[4] and all three contractors derive a significant proportion of their income from abroad (see Table 1.2). They may thus be able to 'shield' their operations from cyclical movements in their national economy and allow more stable long-term funding for innovation. The potential for a fragmented approach to innovation activities by different divisions, however, is avoided by the creation of 'competence centres' and R&D support units in Holzmann and Hochtief to aid dissemination of knowledge and good practice between divisions. Divisional requests for funding are channelled to central co-ordinating groups (that include members of the board of directors) which evaluate the project proposals. To ensure the business orientation of innovative ideas, divisions are expected to fund half the cost of incremental projects originating in that particular division. These mechanisms are in place for supporting small-scale innovations but the central coordinating group also has responsibility for longer-term strategic innovations such as those designed to exploit new markets or which involve inter-divisional funding.

As argued in the section 'Corporate Governance and Innovation in Construction, above, cross-holdings among firms may support long-term relations, which, in turn, may be beneficial to innovation. At all three German contractors, senior managers interviewed regarded competitors, suppliers and clients as the principal source of innovation. For example, Holzmann has collaborative supply chain relations with the electrical engineering and electronics firm Siemens, an executive of which sits on its board. Similarly, Strabag has close relations with the automobile manufacturer Ford also linked to its board.

Cross-holdings may also facilitate national and European collaboration to share technological and management expertise and collaborate on research and development. For example, Holzmann and Hochtief are involved in ENCORD (the European Network of Construction Companies for Research and Development), a European partnership of leading EU construction firms. The strategic objective of ENCORD is to increase awareness of the potential of industry-led R&D by defining common R&D projects, lobbying for the construction industry in the EC and facilitating the exchange of information, best practice and specialist knowledge through seminars and workshops. Similarly, Strabag is involved in SEC (Société Européenne de Construction), a European collaboration, including leading Swedish and British contractors, which aims to share technical management expertise, experience in project financing and in Build, Operate, Own and Transfer BOOT projects and to raise finance for innovation.

Sweden

In Sweden, Germanic corporate governance features have been tempered by an Anglo-Saxon growth in stock market investment following the deregulation of the financial sector in the 1980s. However, 'banking spheres' and family ownership are still predominant in Swedish corporate governance systems.[5] The corporate governance of the four top Swedish contractors interviewed (Skanska, NCC, PEAB and JM) is closer to the Germanic corporate governance typology with strong industry and bank ownership (and also family ownership)[6] and employee representation on the board (see Table 1.3).[7]

The combined interest and influence of banks, family, industrial firms and workers enable the Swedish contractors to invest in firm-specific innovations that demand significant long-term funding. For example, Skanska, with an annual in-house R&D investment expenditure of SEK250 million (in 1998), concentrates in areas considered of strategic importance to the firm. Examples include developments in infrastructure technology, introduction of IT to streamline the construction process, the development of wooden structures, research into the indoor environment and global environmental issues.

In a domestic industry of small size (see Table A.2 in Appendix), the largest contractors have pursued overseas operations through takeovers and acquisitions. As the Swedish construction industry faltered during the 1990s, Skanska and NCC were able to grow considerably in terms of sales and total assets by overseas expansion. At the end of the 1980s, the proportion of turnover Skanska derived from abroad was just 8 per cent. Skanska engaged in an aggressive internationalization strategy culminating in 65 per cent of total annual turnover originating outside Sweden in 1997. The firm's turnover derived from the USA now accounts for a larger percentage of total turnover than that derived from its domestic activities. Similarly, NCC has grown over recent years by expanding significantly its operations in Europe, operating through wholly-owned subsidiaries in Denmark, Norway, Germany and Poland (NCC Danemark, NCC Eeg-Henriksen, NCC Siab, NCC Puolimatka and NCC Polska, respectively). This presence across international markets may explain why NCC and especially Skanska have managed to maintain high dividend payments and high investments in the 1990s. Also, it may enable stable long-term funding for innovation.

In common with the German contractors, Swedish contractors are not only decentralized geographically but also by business area.[8] And, again despite extensive decentralization of management structures, the Swedish contractors have avoided a fragmented approach to innovation activities. Similarly to the German contractor's 'competence centres', contractors in Sweden have established mechanisms to collect and disseminate technical information across their decentralized structures. For example, in Skanska, most R&D

Table 1.3   Swedish contractors: ownership, control and structure

Company	Leading shareholders (% of voting shares)	Controlling interest	Income from abroad (%)	Active European holdings	Active overseas holdings
Skanska	Mutual bank (13); foreign individual (11); bank sphere (9); state pension (7)	Bank sphere (35); family (10); mutual bank (8); foreign S/H (6)	65	16	9
Nordic Construction Company (NCC)	Family (33); family (13); bank sphere (13)	Family (48); family (16); bank sphere (10)	40	9	0
PEAB	Foreign institution (23); family (19)	Family (60); foreign institution (10)	15	7	0
JM Byggnads och Fastignets AB	Skanska (27); mutual bank (10); bank (6)	Skanska (45); mutual bank (7)	0	1	1

Notes: See notes to Table 1.2.

Source: Individual firms' annual reports (1998), European International Contractors (1998).

work is conducted centrally, and is co-ordinated by Skanska Teknik. Skanska Teknik integrates the firm's technical expertise and disseminates knowledge and experience across the firm. It also supplies the business areas with consulting services in selected fields of technology. Moreover, research funding is available for technological development related to each division's core activity. Each division of the firm has a separate R&D budget, a different R&D focus and R&D manager. In addition, developments related to products and processes implemented in major projects are carried out in the construction and industrial divisions in Sweden, Finland and Denmark. Innovations introduced in the different geographical divisions are diffused between divisions through exchange of personnel. Skanska also looks beyond the boundaries of the firm, engaging in domestic and EC-financed research and development projects to broaden its general technical knowledge and competence.[9]

In NCC, despite the group's enlargement through mergers and acquisition, R&D activity is concentrated within the firm's central R&D unit. R&D with long-term strategic aims and R&D with a group-wide interest are managed and co-ordinated through the group's collective R&D resources. R&D activities prioritize co-operation with technical colleges, participation in national and international research programmes and co-operation between firms within the group.[10] NCC Technology plays a central role in the development, application and dissemination of technical knowledge and skills within the firm. NCC Technology has 140 specialists across disciplines such as project planning, project management and technical development in the construction, civil engineering and installation areas. The unit offers technological expertise (systems know-how, leading-edge expertise and technology and process integration) across product areas based on advanced understanding of the construction process generated through close co-operation with NCC's production operations.[11]

As argued by senior staff at Skanska and NCC, the importance of international competitiveness has increased the need for technical expertise within firms, at the expense of economic and legal experts. In this context, links to universities are important. Skanska and NCC have staff working in universities on projects connected to in-house R&D. As we will see in the next chapter, both firms support postgraduate students. There is no assumption that postgraduate students will develop innovations that will be implemented in the firm, the idea is to develop a broad knowledge pool and a network of contacts.

As in Germany, cross-holdings and the fact that senior executives sit on the boards of many industrial firms have enabled long-term relations with suppliers and customers. For instance, NCC has collaborative links with Ericsson, the largest supplier of mobile communication systems, for research into the application of telecommunications in intelligent buildings. Similarly,

PEAB has been working with Ericsson to create a system of bar codes and data transfer between contractors, materials suppliers and materials producers. Also, Skanska has been innovative in the housing sector, co-operating with IKEA (the largest furniture retailer in the world) in building cheap wooden frame housing ('Bo Klok' or Live Smart) (see more on this in Chapter 2). NCC has formal alliances for the development of materials with thermal insulation firms such as Gullfiber (the leading manufacturer of mineral wool in Sweden) and plasterwork suppliers.

Finally, the Swedish contractors tend to be involved in international collaborations. For example, to engage in major infrastructure projects in Europe, NCC has joined the strategic alliance SEC and Skanska is part of ENCORD. NCC's desire for more international exposure and involvement in major infrastructure projects is demonstrated through its participation in SEC. Also, NCC collaborates with the German contractor Strabag and the Italian contractor Impregilo, undertaking major projects in Southeast Asia. NCC also co-operates with Impregilo in the Russian and Baltic markets.

Denmark

While the Danish corporate governance system corresponds to the Germanic typology, the top Danish contractors included in our research (Hojgaard and Schultz, Monberg and Thorsen, Skanska Jensen and NCC Danemark) are smaller than the German and Swedish contractors. The latter two are in fact wholly-owned subsidiaries of the two largest Swedish contractors.[12] All contractors interviewed have a two-tier board system and have three employee representatives on the board. The CEO and the chairman of the supervisory board cannot be same person. Employers' pensions and retirement schemes are not allowed to have dominant positions in the firm either together or separately.[13]

In NCC and Skanska Jensen, strategic decisions on innovation are taken at the level of the parent firms. Senior managers interviewed at Hojgaard and Schultz and Monberg and Thorsten claim that their firms cannot be regarded as innovative, partly because, owing to their size, they do not have cash flows capable of financing significant R&D on their own. Profits are low, particularly in light of the high building costs and low-margin contracts (see Figure 1.1). Contractors do not derive a large proportion of their income from abroad and their turnover is predominantly influenced by the state of the domestic economy (see Table 1.4). This may make Danish contractors more vulnerable to the cycles of the Danish economy and may therefore give them less stability for the funding of long-term projects.

In our interviews, senior managers argued that the government was the principal source of information and encouragement for the adoption of new

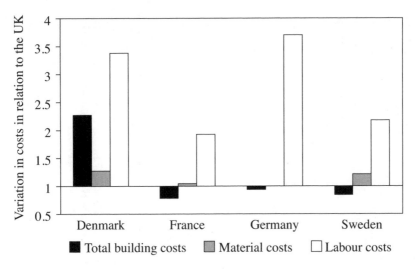

Notes:
1. This figure shows index linked total building, labour and materials costs across the five countries in comparison with the UK (=1). For example, French total building costs are 0.8 times less than the cost of UK building costs; Danish materials costs are 1.3 times greater than UK material costs; and German labour costs are 3.6 times greater than UK labour costs.
2. Comparative materials prices in 1998 index linked to the UK. Materials prices for each country represent average material costs, expressed in £ per unit measurement, across nine essential building materials for construction conducted in capital cities in 1998. Actual UK value = £292.63 per unit of material.
3. Comparative total building costs in 1998 index linked to the UK. Average price of building across seven types of residential and non-residential construction in the capital cities in £ per meter squared per floor in 1998. Actual UK value = £149.6 per metre squared per floor.
4. Comparative labour costs in 1998 expressed in relation to the UK. The labour cost represents the 'all-in rate' which is the gross hourly cost of employing a skilled site operative based on the standard working week of the country and includes insurance, statutory contributions and taxes and is quoted in pound sterling for 1998. Actual UK value = £7.36 per hour.
5. Figures for Germany refer to an average of data for Berlin and Frankfurt.

Source: Costs data from Gardiner and Theobold (1998).

Figure 1.1   Comparative materials, labour and total building costs index linked to the UK

technologies and contractual arrangements between firms. However, it was also apparent that contractors were learning from the experiences initiated by the government and taking a more active stance regarding innovation. The Danish government has promoted collaborations across the supply chain and has financed demonstration projects especially in the field of industrialization. NCC Danemark, Skanska Jensen and Hojgaard and Schultz have been involved in a four-year government-initiated project promoting vertical

Table 1.4 Danish contractors: ownership, control and structure

Company	Leading shareholders	Controlling interest	Income from abroad (%)	Active European holdings	Active overseas holdings
Hojgaard Holding	Family charitable foundation (>50); dispersed ownership (<50)	Family charitable foundation	9	5	0
Monberg & Thorsen	Founder families (35); employees (27); institutional investors (18)	Families; labour market pension; employee pension	13	1	1
Skanska Jensen	Industrial parent (100)	See Skanska	–	–	–
NCC Danemark	Industrial parent (100)	See NCC	0.5	2	0

Notes: See notes to Table 1.2.

* Hojgaard Holding and Monberg and Thorsen Holding are the holding companies for Hojgaard and Schultz and Monberg and Thorsen respectively. The holding companies own 100 per cent of each company and the leading shareholder and controlling interest figures refer to those of the holding companies. Holding companies are used to centralise the provision of financial, managerial and/or marketing functions across subsidiaries.

Source: Individual firms' annual reports (1998), European International Contractors (1998).

collaboration, the Process and Product Development demonstration project between 1994 and 1997 with a total budget of DKK50 million. NCC Danemark's project (Comfort House) focused on developing a light build house (using steel and gypsum) suitable for industrial production, implementing new forms of integrated co-operation between the contractor and the building consultants. Unfortunately, the project ultimately suffered from lack of volume – the initially agreed quota was significantly cut, leaving NCC Danemark with no market for its innovation. Skanska Jensen's project (CASA NOVA) included the development of the first wood-based industrialized system for apartment buildings. The project took advantage of new contractual arrangements and vertical integration, using IT/CAD to facilitate the coordination of the project team.

Hogjaard and Schultz's project focused on process development and vertical collaboration. The design phase of the building project was divided into a number of separate process units (carcass, bathroom, façade, roof and interior fittings), which paid no regard to traditional trade boundaries, but where each process unit could be detailed without intervening in other processes. The use of IT in a common database ensured that all necessary information was available to the parties involved at the right point of the process. This is a good example of an initiative by the government continued by the contractor. Prior to the Process and Product Development programme, there was no market for this type of arrangement, but, more recently, Hojgaard and Schultz has marketed itself as a provider of this type of building process, using it in new projects. The senior managers interviewed at the firm said that although some clients are interested in this type of arrangement, not all clients are well informed and prefer the option of lowest-cost tender. However, with two-thirds of Hojgaard and Schultz's work through repeat contract, there is a high probability that this building process will be used again in future projects.

Programmes such as the Process and Product Development have been important in making the largest contractors recognize the need to be active in innovation. As argued in our interviews with Hojgaard and Schultz, contractors in Denmark still consider innovation in project terms as opposed to more strategic terms. Hojgaard and Schultz is now beginning to regard itself as a 'learning firm', changing its approach to the construction process and prefabricated solutions with a view to repeat business and integrating the experience of projects into its business processes rather than considering each project in isolation.

France

French contractors have a Latin or family-controlled corporate governance system. Despite firms having the choice of either a one-tier or a two-tier board system, 98 per cent of boards are unitary and there is no distinction made between non-executive and executive directors by French law. Conventionally, two-thirds of directors can be non-executives but they tend to represent major shareholders. Directors have wide powers in relation to the management of the firm. Ownership concentration tends to be high and cross-shareholdings, government control and family control are important. In general, there is a stronger emphasis on shareholder's interests than can be seen in Germany, but stakeholders' interests remain the most important. For example, management can be removed by shareholders at will, but an absence of 'one share, one vote' limits the influence of the shareholder on management decision-making. The Latin system is thus in between the Germanic and Anglo-Saxon, while the countervailing influence of employees and of independent shareholders is less than in the Germanic model, the stock market plays a lesser role than in the Anglo-Saxon model.

Table 1.5 includes the details of ownership of the top three French contractors interviewed as part of our research (Bouygues, GTM and SGE). Ownership of Bouygues is family-concentrated and in GTM and SGE it is divided between institutional investors and dispersed ownership. Unitary boards control Bouygues and SGE, while a two-tier board system operates at GTM.[14]

French contractors earn a significant proportion of their income from abroad and operate across Europe and overseas. All three firms have seen strong sales and total assets growth between 1994 and 1998. Increasing (already high) turnover and without a shareholder priority, Bouygues and GTM have been able to increase cash flow reserves and diversify whilst providing consistent or slightly increasing dividends, respectively. SGE, despite strong sales and assets growth over the last five years, diverted resources to restructuring in order to increase its scope for business activity while issuing zero dividends. In general, however, dividends have remained fairly constant over the five-year period despite sales and assets increases and the firms have been able to channel profits toward firm growth and innovation.

Concentrated family ownership may explain the support for firm-specific investments that may demand significant R&D funding. For example, Bouygues has been involved in the development of a number of innovations: bridges using special steel tubes (which have been used in Madagascar); technologies to minimize the effect of waves on sea protection walls (applied in Lebanon); self-compacting concrete without vibrations. Similarly, Campegnon Bernard, the building division of SGE, has been developing high-performance concrete mixes and has been working in the field of monitoring and laying of concrete through self-laying and self-layering concrete formulae. It has undertaken research into the durability properties of ageing

Table 1.5 French contractors: ownership, control and structure

Company	Leading shareholders	Controlling interest	Income from abroad (%)	Active European holdings	Active overseas holdings
Bouygues	Family (16); domestic institutional investor (15); foreign non-bank financial institution (7)	Family; institutional investor; employees	36	23	31
Groupe GTM	Domestic industrial parent (49); dispersed ownership (47);	Domestic industrial parent	43	28	45
SGE	Domestic industrial parent (50); dispersed ownership (44)	Industrial parent	35	22	14

Notes: See notes to Table 1.2.

Source: Individual firms' annual reports (1998), European International Contractors (1998).

concrete and has developed computation software for analysing the dynamic behaviour of structures.

French contractors place special emphasis on knowledge management and the diffusion of knowledge across the firm. For instance, a new board has been established within GTM to deal with strategic R&D. The group includes the top 15 technical managers and the chief executive. Research topics are discussed in terms of the future strategy of GTM, its strengths and weaknesses, the type of services it wants to offer its clients, the strategy regarding supplier relations and the future management of its structural work. Due to the need to appropriate the benefits of R&D, a business plan is developed for each R&D project. Funding though does not necessarily come from the head office. In civil engineering, 90 per cent of funding comes from in-house divisional resources. In the building division, external sources are common, in particular from the government and EC. The parent company of GTM, Suez-Lyonnaise des Eaux, also has an innovation budget of FF20 million to fund 50 per cent of projects if they benefit all group firms.

To diffuse knowledge within the organization and to learn from in-house divisional experience, GTM has developed an intranet database of all GTM projects and holds 'technical days' to enhance interaction and share knowledge between young engineers and experienced managers.[15] Also, GTM grants an internal innovation prize every two years to reward innovation, to collate and record innovations, and to disseminate knowledge and encourage staff to develop further original ideas.

Cross-holdings may explain the high incidence of long-term relationships with competitors and suppliers. For example, Campegnon Bernard is involved in two long-term relationships with suppliers: working with cement producers in the development of new concrete and working in a long-term agreement with a tunnelling and boring machinery supplier to develop a new guiding technology. The two partners share the cost of the research and share the revenues from the patent.

In addition to its participation in ENCORD, GTM has actively sought European research projects, the tenders of which are secured from the EC, which pays 50 per cent of the cost. For example, GTM has been involved as co-ordinator in four European projects over the last few years, collaborating with contractors in Italy, Spain and Germany. As a result of a project carried out between 1992 and 1997, a laser technology for use on historical buildings was developed. The project was so successful that the laser cleaning services are now provided by two subsidiary firms of GTM. Despite leaving ENCORD, Bouygues has maintained its interaction with other contractors working in international collaboration with, for example, the British contractor AMEC and the Spanish contractor Dragados. Bouygues has also sought co-operation with sub-contractors. For example, in a European research project into tunnelling and boring, Bouygues has been working with specialist firms in Italy and Germany that make the tools used on the front of the tunnelling and boring machines. The scheme highlights one of the reasons behind Bouygues' emphasis on internal R&D – the threat of its knowledge being relayed to third parties. Where specific co-operations are sought, for example, between Bouygues and the University of Liege in Belgium, confidentiality papers are signed.

United Kingdom

The Anglo-Saxon model of corporate governance comprises a unitary board with a board of directors made up from top management executive directors and external non-executive directors. The non-executives are further divided into those 'related', for example, a major shareholder or supplier or customer, and those who are 'independent', with no connection to the firm outside their directorship. Most UK boards include a majority of executive directors, but the importance of non-executive directors has increased in light of the 1992 Cadbury Committee Report on the financial aspects of corporate governance.[16]

The three top UK contractors covered here (AMEC, Carillion and John Laing) all have a unitary board system, each with three independent non-executive directors amongst seven or eight executives. There has been a shift toward institutional ownership in the UK, particularly with respect to pension funds and insurance firms, and away from share ownership by individuals.[17] Indeed, institutional investors principally own each of the three top contractors (see Table 1.6).[18]

Because of the particular ownership structure, UK contractors are concerned about the effects of a low share price on clients, staff and their corporate image (see White 2000). Profits are important and dividends are at the forefront of decisions regarding profit allocation. Dividends have increased consistently over the last five years for the three top contractors. Even during the recession between 1990 and 1994, when turnover was falling and profits were negative, dividends were maintained at a constant level. Indeed, in a surprising admission, a senior manager of Carillion, formerly Tarmac's building arm, argued that the extent of R&D funding was not made public because there may be pressure to re-channel it to boost dividends and shareholder value.

Also, senior managers interviewed at UK contractors argued that the level of R&D funding was not the best indicator of innovation. They contended that since they have such a small asset base in comparison with turnover (for example, AMEC's ratio of assets employed to turnover is 9 per cent; Carillion's ratio is 8 per cent, Laing's ratio is 18 per cent),[19] the most important resource for innovation is their project management ability. This view of

Table 1.6 UK contractors: ownership, control and structure

Company	Leading shareholders	Controlling interest	Income from abroad (%)	Active European holdings	Active overseas holdings
AMEC	Insurance institutional investor (12); insurance institutional investor (11)	Fund management; unit trust	39	3	5
Carillion	Fund management (16); fund management (14); insurance institutional investor (11)	Institutional investors	17	4	2
John Laing	Charitable foundation (7); charitable foundation (6)	Director trustee controlling interest	17	2	5

Notes: See notes to Table 1.2.

Source: Individual firms' annual reports (1998), European International Contractors (1998).

innovation explains the reorganization of the larger contractors that occurred during the 1990s. Contractors shifted from investment in productive activities to outsourcing (for example, increasing plant hire), an increased involvement in the management of construction and a strategy of conglomerate (moving, for example, into services and facility management) and multinational diversification (see also Miozzo and Ivory 2000).

Because of this change in business strategy, the top British contractors are involved in developments in project management and supply chain management. For example, AMEC is working with Loughborough University on an Integrated Design and Construction project to further the understanding of the design process in production information. Together, they are developing techniques of engaging the supply chain in value engineering. The aim of the project is to improve interfaces in the supply chain and transfer the benefits of supply chain integration from frequent contracts with large clients to occasional contracts with smaller clients. AMEC is also involved in research projects with Salford University developing ways to ensure the continuity of information through all the business processes involved in building.

The corporate reorganization of the largest contractors has involved the establishment of internal institutions and procedures to determine the direction of their strategic innovation. Traditionally a design management construction contractor, AMEC has developed multidisciplinary teams by separating building specialities (leisure, retail, etc.). These teams include members from all professions (such as architects, engineers, project managers and sales staff) and are responsible for whole projects from design through construction to delivery (including organization and project management, and pre- and post-contract cost-control operating).

Due to their income from operations abroad, property and other sectors, combined with an ability to reduce margins and bid for smaller contracts, large contractors suffered less in the recession of the late 1980s than their smaller and medium-sized counterparts. Indeed, during the 1990s the contractors sought to diversify further to avoid the potential impact of future recessions. For example, AMEC's percentage income from abroad more than doubled between 1994 and 1998 (to 39 per cent), an increase that can be largely attributed to expansion within Europe. Similarly, though not to the same degree, during the same period, Laing's percentage of income from overseas increased from 12 per cent to 17 per cent and Carillion's increased from 14 per cent to 17 per cent.

Despite decentralization, however, UK contractors have made deliberate efforts to diffuse innovations and knowledge within the organization. Laing undertakes audits of its internal procedures, capturing its best practice and disseminating the information through conventional in-house training schemes and lectures, and electronically via the intranet. In Carillion, the technical library and intranet provide a database of in-house and external experts, who have been involved in different projects. Carillion also has a team working in an Integrated Management System which is concerned with the dissemination of information, promoting best practice and measuring performance against established targets (for example, against a score-card and against performance at other sites). In most UK contractors, there are individual prizes for innovation and innovation prizes at different levels of the organization (business group level, engineering services level, and organization level). As argued by Carillion, these facilitate the further transfer of best practice within the organization. Subsequently, each department then records, as part of its performance indicators, savings or improvements that have been generated and these feed into bonuses for senior executives. British contractors also make use of panels or forums, which include internal and external experts, to feed into continuous improvements. For example, directors from all major areas of Carillion and representatives of four key universities meet twice a year in an Innovation Forum. Also, Carillion has been involved in a long-running board-driven initiative to examine all business processes. As a result of Team 2000 (a team of 12 people who examined the firm's processes over two years), procedures within Carillion were established to diffuse and replicate best practice from one area in others.

Links to the government have been important, especially for supporting new contractual collaborations and closer co-operation with clients (see Miozzo and Ivory 2000). For example, Carillion has seconded an executive to the UK government-supported Egan group and currently seconds a senior director to the position of Director of the Construction Best Practice Forum and seconds another executive to the Movement for Innovation (M4I) panel.

IMPLICATIONS AND POLICY RECOMMENDATIONS

Our research findings demonstrate that contractors are important sources, and adopters, of innovations that improve construction technologies and integrate the different activities and innovations introduced by different parties in the construction process. Nevertheless, the role that contractors play in the development and diffusion of innovation differs by country. The explanation of these differences is not simply a function of differences in management approaches to innovation and R&D. In the above illustrations, we have emphasized that the nature of certain features of corporate governance shape the extent to which strategic control is in the hands of those that have the incentives and abilities to invest in innovation. In particular, we focused on differences in ownership, finance, and management structures and the way that these shape the influence that key players (workers, shareholders, banks, families, government, suppliers and customers) may exert on the decisions to allocate resources to uncertain and irreversible investments in innovation. Also, internal mechanisms to diffuse knowledge, inter-firm collaborations and relations with the government and research organizations may explain lower building costs and better construction performance, despite evidence of relatively high wages, in some European countries.

In this chapter, we argue that the development of strategic innovations and operational capabilities depend on the role of three factors:

- the structure of ownership and management of contractors;
- the creation of institutions within the firm to facilitate the diffusion of new processes and practices across the different divisions; and
- long-term relations between firms and collaborations with external sources of knowledge.

We conclude with a brief comparison of contractors from the different countries to illustrate each of these features in turn, and the way they influence strategic investments in innovation. Contractors with a Germanic (or Latin) corporate governance structure combined with high turnover, margins and diversification, are in a good position to develop a long-term strategy of research and development. In this case, the influence of banks, industrial firms and workers ensures financial commitment to uncertain firm-specific investments in innovation. In the absence of this influence, contractors are likely to be pressed into meeting the short-term interests of shareholders, and to meet dividend payments rather than engage in long-term investment in production technology and machinery. Some examples follow:

- in Germany, the combined interest of banks, non-financial firms and workers facilitates the involvement of contractors in long-term research and development (in some cases even in spite of financial difficulties) in areas such as communications and construction materials;
- also, in Sweden, banks and family ownership, together with large cash flows and overseas expansion, allow contractors to have a long-term commitment to R&D and still maintain dividend payments;
- whereas, in the UK, contractors are principally owned by institutional investors and there is a strong pressure to maintain dividends; UK contractors have shifted from investments in production technologies to investment in the management and control of the construction process.

Investment in research and development at a corporate level within large contractors is a necessary, but not sufficient, condition for securing innovation in construction processes and products. Implementation of new technologies is contingent upon the effective restructuring of internal corporate functions to enable dissemination of best practice across the different divisions of the firm, as follows:

- in Sweden, contractors have established very effective internal mechanisms to coordinate innovation and have separate business areas devoted to evaluating, co-funding and disseminating innovative activities across the firm;
- French contractors have developed sophisticated knowledge management practices such as a special board to deal with R&D and innovation prizes and intranet;
- in the UK, firms have forums, audits of innovation, technical libraries, databases of in-house and external experts and innovation prizes; and
- also, in Germany, contractors have established 'competence centres' to facilitate the diffusion of innovation within the firm.

Links with other construction firms and universities assist in the development of innovation, as follows:

- Swedish contractors have external links with universities to develop a knowledge-based approach to innovation and have also developed strong relationships with manufacturing firms and other European contractors that allow them to share knowledge in areas such as innovative housing developments, construction materials and communications; and
- German and French contractors are engaged in collaborations with other firms and in European partnerships to promote learning.

Although ownership and financial features of a country are difficult to reshape, it is evident from our discussion that government can have an important role in guaranteeing public markets for innovative firms. In particular, government can set an example to industry by supporting alternative procurement relations. More importantly, it can act as a broker to bring together collaborations and networks. Government can facilitate relations between contractors and a wide range of institutions such as universities and specialist subcontractors. In this role it can ensure that the benefits of adopting innovations spill over to the weaker organizations in the network including smaller subcontractors and, through encouraging adequate employment protection and training provision, to skilled labour. Thus, innovation among contractors is spurred in countries where government provides financial support for pilot projects, or supports collaborations among construction firms:

- in Denmark, where contractors are smaller in terms of turnover and have cash flows too low to be able to fund significant research and development on their own, government plays an important role in spurring innovative projects through promoting collaborations across the supply chain and in financing demonstration projects. These initiatives have encouraged contractors to take a more active role in innovation; and

- in the UK, the government may play a less explicit role financing innovation projects but, by recommending contractual arrangements and close co-operation with clients, it has shaped innovation patterns.

NOTES

1. External sources of information and knowledge may include universities, research institutes, trade associations, the government, quasi-governmental bodies, private clients, specialist suppliers or subcontractors, professionals, such as architects or engineers, and other domestic and international contractors.

2. For example, as demonstrated by the near collapse of the largest German contractor, Holzmann, in 1999, there remain significant lucrative divisions within the firm that can be sold if the firm fails. While there was little re-sale value on Holzmann's German operations, its American engineering divisions were highly sought after given the boom of the US economy.

3. Domestic banks and reciprocal shareholding are in evidence in terms of block ownership and seats on the board in German contractors. For example, Deutsche Bank directly holds over 20 per cent and sits on the supervisory board of Holzmann; Commerzbank has significant cross-shareholdings in Hochtief through RWE and also sits on the board; and representatives of both banks sit on the board of Strabag.

4. Holzmann is organized into Philip Holzmann Germany and Philip Holzmann Worldwide and Strabag into Strabag Germany, Strabag International and Bau Holding Austria. Holzmann also has a decentralized management structure across competencies (heavy construction, plant engineering and building services, engineering, project development and facility management) within geographical divisions. Similarly, Strabag is organized around competencies within regional divisions. In contrast, Hochtief has consolidated its position as a traditional contractor, organized internally into four divisions (building, civil, airport management and international).

5. Corporate governance systems in Sweden were shaped by the post-First World War crises of the early 1920s. Beginning in the 1930s, banks, unable to hold shares in other firms, retained their influence over industry by switching industrial shares into newly established investment firms and offering shares of the holding companies to bank customers. The 'bank spheres' worked alongside the firms offering them financial security with a long-term perspective (Adolfsson et al., 1999).

6. As shown in Table 1.3, ownership of Skanska lies with banks and foreign shareholders. Control rests with SHB banking sphere that owns over two-thirds of the A shares, and with the Kamprad family (owners of IKEA) who own 20 per cent of the A shares (Skanska's A shares carry ten votes per share while B shares carry only one). NCC is owned and controlled by banks which, at the second tier are ultimately owned by two families (Nordsterjnan-Johnson and Lundberg). Ownership and control of PEAB rests with the founder family (Paulsson), which owns 91 per cent of the A shares giving them 60 per cent of the voting rights and 20 per cent of share capital. JM's parent firm Skanska AB owns 27 per cent of the share capital and controls 45 per cent of votes.

7. For example, Skanska's board of directors includes, in addition to the CEO, six employee representatives and six industry representatives. The industry representatives also sit on a number of other boards, which may facilitate networking among managers of supplier, customer and competitor firms and long-term collaborations between these. This pattern can also be seen at NCC (nine industry and three employee representatives), PEAB (five industry and four employee representatives and JM (five industry and four employee representatives).

8. In 1997, Skanska reorganized its management structures into four business areas plus group staff units, support firms and Skanska Invest. Of the four business areas, three are geographic: Skanska Sweden, Skanska Europe and Skanska USA (the other area is Skanska Teknik). NCC is organized by country and by construction practice. Six country divisions (Sweden, Denmark, Finland, Norway, Germany and Poland) and one international division have six business areas: civil engineering, housing, building, industry, real estate and invest. Not all the six business areas operate in the country divisions. For example, all six business areas operate in Sweden, five of the six (excluding the invest business area) operate in Denmark, Finland, Norway and Poland and only the civil engineering business area operates in the international division. The other contractors in Sweden, PEAB and JM, also have a decentralized structure and have been consolidating their position in Sweden and the Nordic Region. PEAB AB has integrated upstream, acquiring concrete and ballast producers, whilst concentrating on its domestic, Nordic and European operations in residential and non-residential construction, roads, civil engineering and manufacturing. JM, a construction and real estate firm, has built on its core competence in project development of residential and commercial properties, producing more than 50 per cent of the country's new production of residential units in 1998.

9. Skanksa's participation in seven EC-financed research projects will increase the firm's research staff significantly over the next few years. As part of the EC-financed research being undertaken in collaboration with a number of European partners, Skanska is involved in the following projects: Eurosoilstab (soil stabilization), Elsewise (construction process efficiency), Concur (IT application of Elsewise), IPACS (quality of large concrete structures), Solar Power Envelope (building solar heating systems), Asset (fibre composition in infrastructure construction) and Contecvet (life-time of concrete structures).

10. In 1998, NCC had nine researchers working in the 'Highways and Water' programme across Sweden's four technical colleges. NCC also has employees working as professors in the technical colleges, who also sit on various research governing bodies. NCC co-ordinated the international EC financial programme for self-compacting concrete and made large-scale commitments to multi-year national research programmes (for example, Road/Bridge/Tunnel, Competitive Building and IT Construction and Real Estate).

11. For example, the unit's state-of-the-art expertise includes geo-technology, concrete technology and interior environment. Expertise in the integration of project planning and other construction processes has been developed through co-ordination with suppliers and facilitated through IT-based systems such as 3D-CAD, simulation technology and advanced calculation technology.

12. The organizational structure of the subsidiaries of Skanska is determined by the parent firm. Skanska Jensen's principal focus is the Danish market. Skanska Jensen has a decentralized divisional organizational structure to maintain strong local attachments across the country. Skanska Jensen has ten divisions devoted to national and local concerns. National concerns include project development, equipment, construction and large projects and specialities. Local concerns include building divisions operating in specific geographical locations, for example, Building Zealand, Renovation Zealand, Building Funen, Building Jutland Central-South and Building Jutland North. Each division has its own accountable director who sits on Skanska Jensen's board of directors.

13. Ownership and control are differentiated in Denmark with two types of shares, one of which, predominately held by family and foundations, has enhanced voting power (Weimer and Pape, 1999). Whilst Skanska Jensen and NCC Danemark are 100 per cent owned by their Swedish parents, Monberg and Thorsen is a wholly-owned subsidiary of the Monberg and Thorsen Holding A/S group. The construction firm's activity accounts for 86 per cent of the

holding company's turnover. The founder families own 35 per cent of the total capital of the holding group with a sizeable tranche owned by the employee pension funds and institutional investors. Similarly, Hojgaard and Schultz is also owned by a parent holding group, the share capital and voting rights of which are held by a charitable organization, which owns more than 50 per cent of the share capital, corresponding to 70 per cent of the vote.

14. Bouygues' board of directors has 21 members, 19 on six-year tenure contracts appointed from the shareholders meeting and two employee representatives on two-year tenures appointed from members of the supervisory boards of the group's Profit Sharing, Investment and Corporate Savings Plan mutual funds. The board includes eight division and group directors, two family members (including the Chairman), nine industry, cross-holding representatives and two employee representatives. In 1998, SGE recruited two non-French directors to strengthen the European dimension, increasing the number of directors to 16. All members of the board of directors represent interested parties in the company's future, either through direct major capital ownership or cross-holdings. No employee representatives are included on the board of directors. Two boards, a board of directors of 11 members and a management board of 5 members, control Groupe GTM. No employee representatives sit on these boards.

15. Previous 'technical days' topics have included interventions on existing building and structures, the environmental impacts of construction sites, water regulation and sanitation.

16. The part-time nature of the job of non-executive director means that the outsiders know less about the firm than insiders (executive directors). However, non-executives, according to agency theory, are more likely to work in the interests of the shareholder since executives have other self-interested managerial motives. Furthermore, the less risk-averse independent non-executives should promote R&D spending since they are less concerned about money being spent on fruitless projects. However, recent empirical research seems to contradict these two theoretical assertions (see Donaldson and Davis 1994).

17. This trend was initiated by institutional investors taking over family-owned capital in the 1950s and 1960s, and was strengthened by events such as the 1975 oil crisis, a tax system that favoured institutional rather than private ownership, an increased demand for pension provision and long-term savings and a trend for firms to issue equity to fund investment (Mallin 1999).

18. Significant shares in AMEC are owned by a number of insurance companies (institutional investors). Significant preference shares, which give voting rights and additional fixed dividend per year, are owned by a number of fund management firms and unit trusts (other institutional investors). Institutional asset management investors also own significant shares in Carrillion. The only contrast is provided by John Laing, the directors of which have 'trustee interests' in 40 per cent of the firm's ordinary shares. Over 20 per cent of Laing's ordinary shares are owned across five trusts and charitable foundations.

19. Assets employed is the combination of fixed assets and net current assets.

2. Networks and innovation in construction in five European countries

INTRODUCTION

The nature of ties within and between firms and institutions strongly affects their competitive performance (Lazonick 1993, Porter 1990). These ties are of particular relevance in situations that involve uncertainty arising from unforeseeable future contingencies, a high degree of interdependence between firms, or a credible threat of opportunism. In these circumstances, close and stable relations between firms may contribute to operational efficiency by reducing transaction costs and, by facilitating the sharing of information and risk, may also promote dynamic efficiency based on innovation (Deakin and Wilkinson 1998).

The nature of the link between institutional structures and economic performance, however, remains elusive. The construction industry is particularly well suited for the examination of these inter-organizational relations because it can be regarded as an archetypal network system where a coalition of firms and institutions come together on a temporary basis to undertake a project (Gann 2000, Winch 1998). However, many of the problems of the performance of the construction industry seem to stem from inadequate inter-organizational cooperation.

In Chapter 1, we examined the effect on innovation of national differences in firm ownership, finance, organization, management structures and mechanisms to diffuse knowledge within the firm across five European countries. Countries with a 'Germanic' corporate governance system (in which there is a combined influence of banks, industrial firms and workers) not only tend to ensure financial commitment to uncertain firm-specific investments in innovation but also tend to have stronger inter-organizational networks than countries with an 'Anglo-Saxon' structure. The principal question addressed in this chapter is whether in a relatively low-technology, labour-intensive industry, these network relations contribute significantly to differences in competitive advantage.

This chapter focuses on the construction industries in Denmark, France,

Germany, Sweden and the UK. The authors conducted extensive interviews with the top three to four largest contractors in each country and with other organizations, such as architects, engineers, clients, research institutes and government bodies (see Table A.1 and A.3 in the Appendix for the contractors and other organizations interviewed). The chapter examines the nature of the relations of the main contractors across six dimensions:

- relation between contractors and subcontractors or suppliers of materials;
- relation between contractors and the government (for example, in its regulatory role, or in its encouragement of demonstration projects);
- relation between contractors and universities;
- relation between contractors and architects or engineers;
- relation between contractors and clients; and
- international collaborations among contractors.

The effect of these different dimensions on industry performance and innovation is analysed. For that purpose, the first section explores the particular relation between networks and the nature of innovation in the construction industry. The second section describes the results of detailed case-study research on construction networks in each country and the innovative activities facilitated by these inter-organizational relations. A final section discusses the research results and policy implications of this analysis.

NETWORKS AND THE CONSTRUCTION INDUSTRY

The analysis of the economics of innovation in the construction sector poses an important challenge because of its uniqueness in relation to other sectors of the economy. Construction is often regarded as a mature or traditional sector that makes only a minor contribution to its own process or product technology (Pavitt 1984). Indeed, R&D expenditures in the construction sector are extremely low (see Table 2.1). Its peculiarities are often presented as an obstacle to the introduction and diffusion of technological change. The particular nature of the construction product and process – the physical nature of its product; the 'one-off' designs with no prototypes (or lack of 'production function'); the fact that activities take place at the clients' premises; the high susceptibility to economic cycles; and the fact that its product varies in value over time – are argued to conspire against the adoption and development of innovations. More importantly, the organizational structure of the industry, in terms of the separation of design and construction and the growing degree of specialization in production and use of subcontractors are seen as exacerbating the problems of achieving any coherent innovation strategy (Gann 1994, Nam and Tatum 1988).

Table 2.1   *Business expenditure on R&D as a percentage of total construction output in each country, 1991–98*

	1991	1992	1993	1994	1995	1996	1997	1998
Denmark	0.0076	0.0065	0.0057	0.004	0.0025	0.0026	0.0023	0.0007
France	0.0201	0.0212	0.0217	0.020	0.0213	0.0214	0.0280	–
Germany	–	–	–	–	0.0142	0.0164	0.0188	0.0206
Sweden	–	–	–	–	0.0123	–	0.0166	–
UK	0.0586	0.0519	0.0368	0.0346	0.0239	0.0225	0.1006	0.0953

Source: OECD (2000), FIEC (1999).

At the same time, however, we have witnessed an intense interest in networks as particular organizational forms that may facilitate innovation. It is argued that networks may be characterized by a high sense of mutual interest, active participation by all parties and open communication (Castells 1996, Nohria and Eccles 1992). To cope with new challenges (demand for enhanced quality and variability, constant innovation in products and processes, and increased cost of innovation), firms have had to resort to organizational innovations both internally and through their relations with other firms. Among these are relational contracting, networks and strategic alliances (Best 1990, Piore and Sabel 1984, Powell 1990).

The construction process may be regarded as an archetypal network system, since construction projects are planned and executed in the context of inter-organizational decisions, activities and relations. Organizations of varied forms exchange information and know-how, sometimes on an episodic and sometimes on a continuous basis. Although some attention has been paid to particular organizational forms in the construction industry such as the 'quasi-firm', based on a set of stable relations between a general contractor and special trade subcontractors (Eccles 1981), and to the comparative effects of different national contractual systems in the construction industry (Winch 1996, Winch and Campagnac 1995), little attention has been paid to inter-organizational relations supporting innovation.

Industry practitioners recognize the importance of these networks. This became increasingly obvious during the course of this research. For example, the Director of Research and Development (Building) of the French contractor GTM explained to us:

> Innovation (in construction) can result from a global approach, from links with the supply chain, the government and clients, from developing logistics on-site, and from knowledge sharing. Because construction is old it thinks it doesn't have to change much. Because it's organized by project it is almost like a virtual enterprise – working on a job for six months or three years. Sharing technologies with those who you work with can increase, for example, the speed of the project since the project can only be completed at the pace of the slowest worker.

This chapter explores the complex networks of cooperation and association in different countries that arise from the need to coordinate closely complementary but dissimilar activities for production and innovation. In a project-based industry such as construction, firms must rely on the capabilities of other firms to produce innovations and this is facilitated by some degree of continuing cooperation between those concerned with the development of products, processes and designs.

In this chapter, we examine stable relations between parties that allow feedback processes and enable non-routine, flexible behaviour, and risky investments in new product and process technologies. These relations may take different forms, from acquisition of subcontractors by contractors to long-term relations between contractors and subcontractors inducing the latter to assume risks in a narrow specialization in skills and equipment. Also, we examine collaborations with universities and cooperative arrangements between construction firms designed to pool or transfer technology. These arrangements are very different from the traditional cartels and oligopolistic agreements, since they are undertaken to develop a new product or improve a new technology, often under the sponsorship of governments or public agencies. In these cases, collaboration in some areas may be in sharp contrast to competition in other business areas.

NETWORKS AND INNOVATION IN FIVE EUROPEAN COUNTRIES

This section draws on case-study material to illustrate the inter-organizational relations that explain differences in performance and innovation in the construction industry across five European countries. In each country, we identified the leading contractors in the construction industry and a number of professionals, research institutes and government bodies and negotiated access for carrying out interviews with senior managers.

We identify the main relationships of contractors across the different countries, focusing on relations with subcontractors and suppliers, the government, universities, architects and engineers, clients and international collaborations with other contractors and suppliers (see Table 2.2 for a

Table 2.2 *Strength of European contractors' networks and importance for innovation*

Parties	Denmark	Sweden	Germany	France	UK
Subcontractors and suppliers	Very strong (collaboration and some examples of vertical integration)	Strong (vertical integration and long term relations)	Strong (collaboration along the supply stream, project-by-project)	Medium (some vertical integration and collaboration)	Weak (but increasing) (some emphasis on supply chain management as part of new procurement forms)
	Very important for innovation	Important for innovation	Important for innovation	Important for innovation	Not important for innovation
Government	Very strong (many demonstration projects)	Weak national, some EU	Weak national, more EU	Weak–medium, some EU and national	Strong (encourages partnering with clients)
	Very important for innovation	Not important for innovation	Not important for innovation	Not important for innovation	
Universities	Weak	Very strong (staff exchange, placements and recruitment)	Medium	Weak–medium (project-by-project basis)	Medium (forums and advice)
	Not important for innovation	Very important for innovation	Important for innovation	Quite important for innovation	Important for innovation

Architects/ engineers	Strong (architect as 'general adviser') Important for innovation	Weak Not important for innovation	Weak (regional emphasis enforced by regulation) Not important for innovation	Weak Quite important for innovation	Weak (particularly with growth of 'design and build' type procurement) Not important for innovation
Clients	Strong (open book, repeated work) Important for innovation	Very strong (collaboration for innovation and repeated work) Very important for innovation	Weak (but increasing) Not important for innovation	Weak–medium (but increasing) Quite important for innovation	Strong (partnering with large clients) Important for innovation
International contractors and suppliers	Weak Not important for innovation	Medium (international collaborations with contractors, members of ENCORD, SEC) Important for innovation	Medium (international collaboration, members of ENCORD, SEC) Important for innovation	Medium (international collaborations with contractors and suppliers, supported by EU) Important for innovation	Medium (international collaborations with contractors) Not important for innovation

summary). For each country, we assess the impact of these relations on performance and innovation.

Denmark

The government in Denmark plays a particularly strong role in developing and encouraging networks in the construction industry. It has implemented both direct and indirect policies, initially in pursuit of increased productivity through the encouragement of industrialization, and later by funding demonstration projects based on collaboration between different parties in the construction industry. Already in the 1960s, the government, the construction industry, research institutes and clients were cooperating to improve the construction process encouraging modularization and prefabrication with the aim of producing low-cost, high-quality housing (Bertlesen and Nielsen 1999). Since the late 1970s, demonstration projects have been implemented to subsidize new technologies and seek the commercialization of these technologies.

A number of joint initiatives between government (primarily initiated by the Ministry of Housing and Urban Affairs) and industry have promoted inter-organizational collaboration during the 1990s. For instance, as mentioned in Chapter 1, in 1994, a four-year government initiative (Process and Product Development Programme) invited bids from collaborative consortia (including an architect, an engineer, a contractor and a building society). The winning projects used industrialized processes and prefabricated components and improved building logistics, through the use of information technology at the design, planning and control stages (Ministry of Business and Industry and Ministry of Housing and Building 1995). Also, in 1995, another initiative (Project Renovation) aimed at renovating old buildings, was based on the development of new products and processes through closer cooperation between construction firms. In 1999, a third initiative (Project House) was aimed at facilitating cooperation between major parties in the building supply stream to increase productivity through the adoption of successful practices from the manufacturing industry (for example, systematic planning of products and processes, improved logistics, long-term cooperation and supply chain management) (ATV 1999).

The Housing Director of Hojgaard and Schultz, the largest contractor in Denmark, which had been involved in one of the four winning proposals of the Process and Product Development Programme, explained that their aim was to address the problems inherent in the traditional system of procurement and to emphasize closer cooperation among all the parties in the building process. The Housing Director stressed the benefits of closer cooperation: the earlier involvement of subcontractors and materials suppliers optimizes the use of their expert knowledge and experience of materials and designs; subcontractors gain through financial stability (for example, since demand expectations are more reliable, subcontractors can improve production planning); the project facilitates improved logistics, especially through the creation of a generic database; the client's greater involvement throughout the process facilitates adaptation if deemed necessary and, through an open-book pricing mechanism, makes the client more aware of the financial framework; and the repetition of the project leads to costs and time savings in subsequent projects.

Nevertheless, representatives of the Danish Building Research Institute, an independent institution under the Ministry of Housing and Urban Affairs, pointed out that the main problem with demonstration projects is that many of the products, processes and working practices prototyped may not be used again once projects had been completed. This can be explained by the fact that many projects may not be financially viable without state backing. Furthermore, in the case of the Process and Product Development Programme, the transfer of control over social housing from the state to the local authorities reduced the initially planned number of houses to be built. For example, the Head of Building Renovation in NCC Danemark argued that the Ministry of Housing initially guaranteed the building of 300 housing units between 1994 and 1998. By 1999, NCC Danemark was only required to build 150 units, despite the fact that the firm had designed an innovative light-build house using steel and gypsum for the purpose of the programme.

In our interviews, senior managers from the leading contractors stressed the importance of close collaboration with suppliers of materials and components. Some firms have resorted to vertical integration, as in the case of NCC Danemark and Skanska DK, the first and second largest contractors in Denmark. For example, Skanska DK has internalized materials producers and suppliers (such as prefabricated components producers and steel suppliers) and specialist subcontractors (such as carpenters, electricians, plumbers and bricklayers) and is actively involved in most stages of the building process except specific design and engineering expertise and finishing trades. In the majority of contracts, and unless the client demands otherwise, Skanska DK uses its own bricklayers or electricians. Similarly, the number of sub-contractors used by NCC Danemark has fallen and the contractor conducts approximately 40 per cent of its work with in-house tradesmen. However, as argued by our interviewees, the number of in-house tradesmen is difficult to optimize since some organizational slack must be maintained. The 'hold-up' problem is particularly acute for some building components. For example, there are only a small number of firms manufacturing prefabricated components in Denmark and the price and delivery times fluctuate with varying demand. NCC Danemark has solved this problem by buying a manufacturer of prefabricated components. Although neither the contractor nor the supplier are exclusive buyers or sellers, the contractor remains in control of the delivery terms (moving towards 'just-in-time' supply of components) and can rely on a guaranteed supply, constant price and trusted quality from a known source.

Danish contractors also seek repeated work with their largest clients. For example, NCC Danemark built approximately 75 per cent of the buildings of the Danish beer producer Carlsberg. With an open-book policy, repeated work with clients can be achieved without conventional lowest cost tenders. Contractors are investing in ways of improving their relationship with clients through the use of information technology during the design and building process. However, with the exception of a few firms, clients lack the size and experience to be more involved in the primary stages and instead offer the architect and engineer only a brief description of the building, rather than specifying its required function. In our interviews, contractors argued that there is an important leadership role for government in setting a precedent for clients.

Architects and engineers play an important role in Danish construction. As argued by a senior partner of an architectural practice, this may be because of the smaller size of the main contractors or because of the strong craft traditions in Denmark. The architect is involved in the whole building process, acting as 'general adviser', complementing the specific knowledge of the engineer in most traditional contracts. Architects and engineers have also managed to maintain their strong influence over public building by, for example, persuading the Ministry of Housing to adopt an architectural policy in public contracts. The architect also plays an important part as the interface with materials producers and suppliers, integrating new materials into relevant projects. Our interviews also highlighted cooperation between architects and materials and components suppliers to pool resources and share risks in innovation. For example an architect interviewed had designed a new type of window that was manufactured by a materials producer, dividing the risk between the architect who introduces the window into the design and the supplier who takes on the risk of creating a market for the product.

Denmark is therefore characterized by very strong network relations with government, subcontractors and suppliers, architects and clients. The government has given encouragement to collaborative relations along the building supply stream, with the objective of introducing innovations in construction processes and products. Also, architects are actively integrated into the building supply stream and have an important role in the design and introduction of new products.

Sweden

As in Denmark, the construction industry is characterized by strong networks. In our interviews, the Director General of the Swedish Council for Building Research (BFS) argued that deregulation in the mid-1990s brought a change in strategy of the largest Swedish contractors. Before deregulation, the Swedish construction industry was subject to strict building standards, requiring the contractors to employ large numbers of economic and legal experts. With deregulation, detailed rules in the building regulations were replaced with more functional specifications emphasizing the performance of the building as a whole (for example, overall energy efficiency, indoor climate and security of the building). An important effect of these changes was a transformation in the employment patterns within the two largest contractors, Skanska and NCC, as the need for technical experts outgrew the need for economic and legal advisers. Hiring technically qualified and experienced staff or training incumbent staff became a priority for Swedish contractors. Thus, relations with universities became very important.

Senior managers of Swedish contractors acknowledge the complementary but very important role of universities as a source of specialist knowledge. Collaboration between contractors and universities is conducted through staff exchange, student placements and student recruitment. The Vice-President of Skanska Teknik stated that the firm has a number of staff with placements in universities, working on in-house projects with a rigorous programme to develop new technology related to their business area. Skanska also has 23 students who work four days a week at Swedish universities and one day inside the firm. Similarly, the Technical Director at NCC Teknik stated that the firm has a dozen postgraduate students working on the development of basic construction technologies. There is no expectation that all innovations developed by PhD students will be implemented later by the firm; instead the strategy is to develop a general knowledge pool comprising a group of experts with firm- and industry-specific knowledge. NCC also has a number of studentships co-funded by the state government. For example, in 1999 NCC was engaged in co-funded projects on 'Competitive Building', 'Roads, Dams and Tunnels' and 'Sustainable Buildings', engaging 29 PhD students. Contractors also give financial support to research institutions such as the Development Fund of the Swedish Construction Industry (SBUF) which is funded directly by construction firms in the industry. Contractors argue that it is worthwhile to contribute financially to research institutes the aim of which is to improve the construction industry and the built environment as a whole, since they not only generate generic knowledge but may generate and diffuse knowledge and practices directly relevant to certain aspects of their existing activities.

Swedish contractors collaborate with foreign contractors through ENCORD (European Network of Construction Companies for Research and Development) and SEC (Société Européenne de Construction). ENCORD is a cooperative enterprise between the largest EU construction firms and includes the Swedish contractor Skanska, the German contractors Holzmann and Hochtief, and, until recently, the French contractor Bouygues. ENCORD has three priorities: to define common R&D projects (for example, recycling construction materials), to lobby for the construction industry at the EU Commission and to exchange information, best practice and specialist knowledge to improve the competitiveness of all partner firms. Since all member firms are willing to share information, this international network facilitates knowledge exchange (through seminars and workshops) and is a good source of innovation. Similarly, SEC, a European collaboration including the Swedish contractor NCC, the German contractor Strabag, and the UK contractor Laing aims to cooperate on projects, share knowledge and exchange personnel. Through the three companies, SEC can draw together technical management expertise, huge financial resources and experience in alternative procurement and project financing contracts.

Contractors are vertically integrated or seek long-term collaboration with materials suppliers. The major contractors own concrete and prefabricated concrete suppliers, asphalt plants, gravel suppliers and window manufacturers. NCC has an Industry Business Area, supplying crushed products, asphalt products, ready-mixed concrete, machinery-rental services and engineering services. NCC invests in these divisions and uses their sector-specific knowledge. For example, NCC has been involved in a project on self-compacting concrete with the aim of internalizing not only the production of concrete but also the casting of the concrete into the final structure.

Despite the high degree of vertical integration, due to intense price competition, work is subcontracted out on a lowest cost tender basis and not necessarily done by integrated suppliers. In certain circumstances, where a supplier is providing an asset-specific product or service, or where there is repeated business between the two, contractors may enter into a more formal long-term agreement with subcontractors or suppliers. These long-term contracts are drawn up at a corporate level and may cause conflict within the organization if the project management also has its own preferred suppliers. The benefits however can be seen, for example, with JM, the fourth largest contractor in Sweden, having a three-year contract with Kune, a Finnish supplier of elevators which establishes a fixed price for the elevators, facilitating JM's and Kune's financial certainty, contributing to more efficient planning and delivery and facilitating cost savings from administration.

Similarly, NCC has a three-year contract in the Stockholm area with Sigvard Carlsson, a supplier of building components. As argued by the

Technical Director, NCC has also been trying to decrease the number of suppliers it deals with, identifying strategic producers to develop closer co-operation, especially regarding technical development. These relationships are usually in the form of formal alliances using long-term contracts. For example, NCC has formal alliances with the thermal insulation firm Gullfiber and plasterwork suppliers as mentioned in the previous chapter. NCC also collaborates with suppliers in research and development. NCC collaborates with the Swedish telecommunications equipment supplier firm Ericsson to determine how telecommunications can be incorporated into intelligent buildings.

As argued by the partner of an architectural practice in Stockholm, while contractors and suppliers are collaborating more closely, architects have lost power in the supply stream. The role of the architect has been weakened by the prevalence of 'design and build' type projects, where the contractor has adopted the role of project manager, increasing its capacity to influence the design and selection of materials and marginalizing the architect to questions of design. In these arrangements, architects are procured in the market on a project-by-project basis. Swedish architects and consultants are less powerful today than in the 1960s and 1970s. In particular, the 1990s recession which led to a cut in housing subsidies, a rise in VAT and increasing building taxes on materials and fees, impacted hard on the level of architectural fees, output and employment. Although contractors maintain good relations with some particular architects, the Technical Director of NCC argued that the selection of architects generally depends on the location of the building project, since architects are employed on the strength of their local knowledge and connections. Contractors have lists of key architects for certain types of building and certain geographical areas and although there remains the possibility of repeated contracts, there is little opportunity for long-term relationships. From the point of view of the architect, the short-term one-off nature of the relationship does not encourage involvement in innovative activities, particularly when there is little autonomy in the design stage and when the choice of materials depends largely on cost. According to a senior manager of a contractor interviewed, contractors are only too keen to remove the architect from the project as soon as possible because 'architects are full of crazy ideas to decrease our profitability'.

A similar analysis may apply to the role of consulting engineers (including mechanical, electrical, building services, structural and civil engineers), although their role has been less dramatically affected. NCC and Skanska employ over 300 engineers between them, and because of their size and links with the universities they are likely to attract most of the best engineering graduates through high salaries and benefits and enhanced career opportunities. However, because projects are procured on lowest cost tender there is still a large market for the services of consulting engineers, particularly those with specialist interests or those who work in specific locations. For example, Skanska, on average, employs its own in-house engineers to work on only around half of its contracts.

The major contractors have very strong collaborations with clients. For example, Skanska has partnerships with clients such as the electrical and engineering multinational ABB, the furniture retailer IKEA and the telecommunications equipment producer Ericsson. When ABB requested a new power plant to be built in two years, Skanska representatives argued that it could build it in six months by using a different construction process. Following this project, Skanska has collaborated with ABB on repeated contracts. Also, as mentioned in Chapter 1, it has collaborated with IKEA in building cheap wooden frame houses for small families – 'Bo Klok' (Live Smart). Following the recommendations of a study team sent to the United States to look at the wooden frame industry, Skansa designed experimental buildings using wooden structures, capitalizing on extensions in the building regulations which allow three- and four-storey buildings to be made of wood (in the past it was only two storeys). The idea behind the collaboration was to build low-priced residential flats, which included in the price the services of an interior decorator and SEK3000 worth of furniture. In addition, IKANO, a bank part owned by IKEA and the Kamprad Family, offers loans for up to 80 per cent of the deposit on a Bo Klok apartment. Skanska also underwrite the development, guaranteeing the finance of the tenant association for seven years regardless of whether some flats remain unsold or tenants move.

As in Denmark, the construction industry in Sweden is therefore characterized by strong inter-organizational relationships. Close and stable relationships between contractors and universities and between contractors and materials suppliers may establish the basis for successful performance and process and product innovation.

Germany

Networks in Germany are weaker than in the Danish or Swedish construction industry. There is, however, emphasis on collaboration along the supply stream. As demonstrated by a recent survey, German contractors regard other contractors and suppliers of materials and components as the main external sources of innovation (see also Cleff and Cleff 1999). In contrast to the Swedish case, the leading contractors in Germany regard collaborations with universities and research institutes as of less importance in terms of innovation. Nevertheless, contractors are involved in collaborative projects with universities, and technical universities in particular. For example, Hochtief, the second largest contractor in Germany, collaborates with the

University of Berlin and research institutes on research into design, software, materials and robotics. Top-tier staff from both Holzmann, the largest contractor, and Hochtief teach at the universities to facilitate closer co-operation with the universities and to gain access to the best students. PhD students are sponsored by Holzmann but only after working in the firm for a number of years.

Also, in contrast to Denmark, contractors in Germany believe that the government does not play an important role in improving the performance of the construction industry. For example, neither state nor local public contracts require the contractors to use innovative products and processes. Also, the construction industry was not included in the German central governments R&D funding schedule for 1998 (BMBF 1999).

The largest contractors increasingly rely on long-term collaboration along the building supply stream, with only a few examples of upstream integration. Although the German contractors still consider themselves as mainly builders (in their own words, 'hard hats and hammers'), their involvement in the management of the construction process and preference for alternative procurement contracts is shifting their focus. Until recently the contractors had not actively sought long-term links with subcontractors or suppliers and tended to rely on lowest cost tender procurement. More recently, Hochtief has entered into long-term arrangements with suppliers, including, for example, a manufacturer of lifts. Similarly, the largest German contractor, Holzmann, has closer relations with a number of key suppliers and subcontractors and has begun to implement procurement strategies with suppliers based on qualitative criteria such as past experience, completion times and quality.

Architects have a particular role in the construction process in Germany.[1] The architectural profession is very regional, making it difficult to work on repeated contracts with large firms that operate nationally. This, in part, accounts for the relatively weak links between contractors and architects. An exception is the third largest contractor in Germany, Strabag, which, according to the Director of Business Development, has some long-term agreements with architects and engineers and is active in attempts to maintain the same construction team in repeated projects. This firm is a special case, however, since it has made special efforts to develop collaborative links with all parties through implementation of its 'Guaranteed Maximum Price' contracts, in which the firm is in control of the management of the architects, engineers, subcontractors and suppliers.[2] It has also developed a new production philosophy, 'Who Shares Wins', bringing all parties together at an early stage and maintaining links throughout the lifetime of the building. By contrast, the head of R&D of Holzmann argued that the firm has few long-term contracts with architects and engineers; most are employed on a project-by-project basis.

Long-term partnering with clients is still in its infancy in Germany. Again, Strabag seems to be an exception. Strabag has been involved in an extended partnership with the automobile manufacturer Ford for five years. A top-tier director of Strabag sits on one of Ford's supplier councils for innovation – a council comprising of the top 150 global suppliers – and through meetings conducted twice a year, innovative ideas to reduce costs in both firms are discussed. For example, the above-mentioned 'Who Shares Wins' concept has been discussed in terms of reducing costs and adding value, improving quality and shortening timescales, and creating a safer and healthier working environment for both Ford and Strabag.

Although there is little co-operation between domestic contractors, there is collaboration with foreign contractors through ENCORD (see above), which includes Holzmann and Hochtief, and through SEC (see above), which includes Strabag. These collaborations enable knowledge exchange and the definition of joint projects.

Networks in the German construction industry are not as strong as in Denmark or in Sweden. There are, however, some examples of collaboration with subcontractors and materials producers to improve construction products and processes.

France

In France, most of the working relationships entered into by construction firms are of a weak, non-collaborative nature. The main exception is some international collaboration between contractors and with machine producers to develop new technologies. According to senior managers of organizations interviewed, the weakness of ties – exacerbated by the prevalence of the lowest-cost tender procurement – is one of the most significant barriers to innovation in France.

In France, the government is not regarded as an important source of innovation in construction. According to the Director of Research and Development at GTM, the second largest contractor in France in the area of housing, this can be attributed in part to the structure of the original independent building agency in France (the PUCA), established in 1975 to promote innovation in housing. Of an annual budget of FF100 million (10 per cent is devoted to seminars and dissemination and 50 per cent to research) 40 per cent is channelled toward an experimental housing sector where contractors propose innovative ideas to a jury, which in turn selects projects to be implemented for social housing associations. However, many projects endorsed by the jury never get built because the PUCA cannot negotiate adequate contracts with social housing organizations.

The government could play a potentially important role as a major client, given the importance of public ownership (such as in French Railways, Electricity, Paris Transportation) but these all procure on lowest-cost tender. According to the Director of Research and Development at GTM, 95 per cent of all work is still procured under lowest-cost tender. Only a few contracts are procured by alternative methods, such as 'global cost' projects, in which firms are not merely judged on the building costs but also on their previous performance. There is however a national innovation scheme (the National Research Project) to encourage new technologies in construction. Through this scheme, the government funds 20 per cent of the cost and the remainder comes from firms, laboratories and universities. Although the largest French contractor Bouygues is the head of one of the national groups, its Managing Director does not consider these collaborative projects to be successful in spurring innovation.

The leading French contractors have many links with universities and research centres across France and Europe, though there are few formal links and relations tend to be forged on a project-by-project basis. Universities are often included in government-sponsored projects and national development projects. Bouygues, for example, has developed a new concept of active structure control with universities in Belgium, Spain and Italy. Bouygues attempts to maintain most high-profile research within the firm because of fears of leaks of knowledge. Important collaborations, such as that between Bouygues and the University of Liège in Belgium for instance, require confidentiality agreements to be signed, as mentioned in Chapter 1.

Collaborations among contractors at the national and international level are of some importance. GTM is a member of ENCORD, replacing Bouygues which resigned from the forum because the management staff argued that the firm was getting less from its participation than other firms were. Bouygues' management staff still believe that collaborations with other contractors are a good source of innovation and they have cooperated with AMEC in the UK and Dragados in Spain, without being part of ENCORD. The third largest contractor, SGE Campegnon Bernard, also recognizes the value of working together with competitors and has engaged in recent joint ventures with Bouygues, for example, on the Normandy Bridge and Stade de France construction. Similarly, SGE has been involved in a recent project with GTM developing a new type of composite material.

The European Union (EU) provides additional funding for innovation for contractors and some collaborative projects supported by the EU have been very successful. For example, during the 1990s, GTM was involved as coordinator in four European projects, in which the EU contributed 50 per cent toward total cost. In one of the projects, BRITE-EURAM (1992–97), GTM developed lasers for restoration work on historical buildings. Subsequently, the laser technology has been developed and marketed in-house through two newly established subsidiary firms.³ EU research projects also give contractors the opportunity to develop niche skills in demonstration projects. For example, a project on self-compacting concrete gave GTM experience of site management during the refurbishment of inhabited social housing. Involvement in this project not only allowed GTM to expand its Paris-based operations but also to expand them over the rest of France. In addition, the involvement in European research projects enables contractors to work with Italian, Spanish and German construction firms and universities and research institutes. Collaboration with materials producers and suppliers is also facilitated through EU projects. For example, in a European research project Boygues established a collaboration with specialist tunnelling and boring machinery firms in Italy and Germany.

With little capital investment, there are very few examples of contractors integrating vertically to internalize suppliers or materials producers. Bouygues prefers to enter into long-term links with firms as opposed to acquiring them, particularly when operating in a new geographical area. For example, in Austria, Hungary, the former Yugoslavia, Germany, Portugal and other countries in Europe, Bouygues has developed strategic partnerships with smaller regional contractors and building consultants. Bouygues' construction division only owns one building contractor plus a manufacturer of pre-cast concrete and a number of electrical subsidiaries. SGE owns several specialist firms (for example, specializing in earthworks) but owns no materials suppliers (suppliers are worked with on a project-by-project lowest-cost tender basis). GTM has one quarry and two prefabrication plants and has recently bought a small firm, operating in Spain and France, that specializes in laminated structures for buildings and bridges. GTM management staff considered that by integrating the firm, it would gain additional technical skills and be able to use wood in more projects. Under group ownership co-operation is common. Freyssinet, SGE's civil engineering specialist division, and GTIE, SGE's amalgamation of firms with expertise in electrical engineering and works, are cooperating on the development of a remote control monitoring system for civil engineering works. In addition, research into noise reduction was carried out on three fronts within SGE: Eurovia, SGE's road division, is developing special mixers of concrete asphalt, Sophianne, SGE's Thermal and Mechanical Division, is developing anti-noise techniques and Freyssinet is developing anti-noise road joints.

Long-term contractual collaborations with materials producers and suppliers are rare. Exceptions include the case of SGE, which is involved in a long-term agreement with a tunnelling and boring machines producer. The firms share the cost of the development and the patent revenues. GTM has a small number of agreements with suppliers with which they have worked on development projects. For example, under an exclusive agreement, GTM and a glue supplier co-developed a new process of reinforced concrete using a glue composite.

The role of the architect is less important in France than in Germany or the UK since although architects are involved in the conceptual design of buildings, they have no role in the engineer's design and management (Huru 1992). Meanwhile, French engineers provide a combination of civil, structural and mechanical engineering knowledge, are involved in every aspect of construction design and management and are, more often than not, employed directly by the contractor. The contractors have few long-term links with architects. For example, while SGE employs some architect technicians to demonstrate ideas no architects are employed within the firm. Only in very particular circumstances is the same architect employed in repeated contracts. For instance, SGE hired the same architect for the building of the Stade de France and the Istanbul stadium and also has regular work with bridge architects. (However, this is because there are very few bridge architects.)

Despite the prevalence of lowest-cost tender contracts there is an increasing number of alternative contracts and partnership arrangements between contractors and clients. For example, GTM has an innovative contract with London's Heathrow Airport that allows profit-sharing in the event of cost savings. Partnerships, using contracts such as these, are sought after by the French contractors but French clients do not yet fully support the idea. Bouygues' building division is attempting to build up these types of relationships with French hotel chains. SGE has undertaken several contracts with the same client but prices are negotiated for each specific project. When working in a good relationship with the same client a series of contracts can be agreed. For example, with Hilton, a special cooperative agreement engages SGE to build hotels in certain European cities such as Berlin. However, again, all hotels are negotiated separately and a lowest price is worked out. GTM has other long-term relations with clients. It has used its technological advanced approach to non-residential building (a new type of flooring structure – a mixture of steel and concrete – that allows reduction of the thickness of the flooring and greater span) to secure repeated contracts with Capital and Continental (a US promoter). Having built previously for the UK retailer Marks and Spencer in Paris, the GTM Paris management team went to Marseilles to supervise local builders for another Marks and Spencer store. Under these partnership-type agreements, it is easier to integrate new technologies. For example, with Marks and Spencer, GTM was selected at the preparation phase despite being the highest bidder because using its knowledge of previous work, it was able to implement the logistical, safety and security measures necessary to keep one part of the store open whilst renovating another part.

In France, therefore, leading contractors engage in international collaborations among contractors and with machinery producers to develop new technologies. Nevertheless, inter-organizational relations are weaker than in the other countries analysed.

United Kingdom

In the UK, collaborations between contractors and clients and an emphasis on process management have taken pre-eminence over collaborations with other parties along the building supply stream. Perhaps the single most important change in UK construction over the last ten years has been the leading contractors' shift in strategic emphasis towards the management of the construction process (for example, management planning, management systems integration and management control) and towards the development of contractual arrangements to improve the delivery of the product to the client (for example, target costs, alliances and partnerships). For example, the seventh largest contractor in the UK, Laing, now operates as a single entity (there is no longer a Laing Civil Engineering Division, a Laing Building Division, etc.). Laing will commission any type of project from any sector. As part of the restructuring, through a process of 'category management', Laing has moved toward forming strategic alliances with design organizations in certain specialized sectors of the market. Thus, the contractors have had to develop front-end skills, enabling them to add value to the project by managing the process better through improved client understanding, contractor and designer skill assessment and the consideration of whole-life costing – multidisciplinary skills, previously disregarded in the industry.

Indeed, in the engineering, building and construction divisions of the fourth largest contractor in the UK, Carillion, between 60 and 70 per cent of current work is whole-life costing, target costing, partnerships or some kind of alliance. Under these procurement forms, profit margins (and risks) are higher for contractors. The building division works closely with 'key accounts' such as the UK supermarkets ASDA and Sainsbury's, the retailer pharmacy Boots and the retailer Marks and Spencer, negotiating on all aspects rather than solely on the financial one. Through close cooperation with clients such as the UK Highways Agency, the Environment Agency and Railtrack and open-book target-cost contracts, contractors argue that they can achieve high added-value engineering solutions. Also, senior staff at Carillion argue that these relations facilitate higher profits (while normal contracts generate profits of approximately 2.6 to 2.8 per cent, for road and railway maintenance contracts, margins are between 8 and 12 per cent). In a surprising revelation, a senior Carillion executive argued that open-book contracts are suitable since 'the more enlightened clients need to see the contractors making a profit'.

The public sector's direct involvement in the construction industry has diminished in the UK, particularly through the introduction of the Private Finance Initiative (PFI), with construction firms often financing, designing, building and managing the public sector facilities in the long term. PFI, however, has come under attack from public policy researchers and the media. PFI developers like Carillion have renegotiated interest payments on loans they take out to build and operate prisons or schools but continue to receive the same payment from the public sector. In other words, contractors are able to refinance the loan after the construction phase, when the highest risks are past, but do not allow the government to renegotiate the terms of the contract. Critics argue that the risks are often overstated, creating unjustified windfalls and diverting public funds to private corporations increasing their involvement in public services (*The Observer*, 8 July 2001).

Emphasis is placed by contractors on collaborative relations with universities. Money channelled towards universities does not carry high opportunity costs since the contractors must provide a business case for the funding. In addition to being the best place to test prototypes, universities are often represented on innovation forums or panels within the largest contractors and senior executives from the top contractors sit on university advisory committees. Government funding bodies, such as the UK Engineering and Physical Sciences Research Council (EPSRC), support high-cost in-house construction firms' projects. Contractors contribute to pools of industry collaborative finance (through, for example, the Building Research Establishment and the Construction Industry Research and Information Association) as a means to keep abreast of the developments of their competitors and technological changes. Carillion, for instance, seconded senior managers to policy bodies such as the Egan group, the Movement for Innovation Panel and Construction Best Practice Forum.

Collaboration between contractors and clients has been regarded as the main way of improving the performance of the construction industry in the UK. It would be easy to associate increased partnering between contractors and clients with the UK government initiatives of the mid- to late-1990s (especially the Latham Report of 1994 and the Egan Report of 1998) to encourage more collaboration between actors in the building chain. But, despite their importance in adding impetus to changing the tide, the transformation in British contractors began independently.[4] Partnering agreements tend to be with the largest clients such as the largest supermarket retailers, leading chain stores, government departments and firms previously in the public sector.

To achieve real benefits, however, these arrangements ought to lead to repeated projects. This explains why Laing attempts to transfer some of its process management skills to clients. For example, Laing recently developed a data management manual and worked with clients to assist them in integrating 'value added process' into all their projects. Therefore, even in cases where Laing is not part of the inception of the projection (for example, if the project was speculative or the client acquired a site independently), the client knows that Laing could be included later or in subsequent projects since it will be experienced with its procedures.

Contractors acknowledge that whilst new forms of procurement have created a closer relationship between contractors and clients, they have done little to address the problems created by the fragmented building supply stream. However, initiatives between the contractors, universities and others in the supply chain have started to examine this problem. For example, Carillion's engineering and construction division, in which most business is conducted with 20 key suppliers, has undertaken a '360-degree appraisal' with their key suppliers to identify areas in which the supplier is under-performing and to see whether Carillion can support the supplier's needs. In addition to this appraisal, the supply chain relies on the concept of mutual dependency, where all actors become involved in sharing risk and working towards the same target. The concept of mutual dependency is particularly important when considering sustainable materials and components, where it is often appropriate to bring the supplier in at an early stage, for example, to assure the client of the source of timber or other materials.

In the UK, inter-organizational networks in the construction industry are weaker than in the other countries analysed. The only very strong collaboration is that between contractors and clients, which has received added support from the government.

IMPLICATIONS AND CONCLUSIONS

Although the construction industry can be characterized as a relatively low technology industry, we find significant differences in productivity between countries. Denmark's productivity has consistently doubled that of the UK between 1991 and 1999. From an examination of Figures 2.1, 2.2, 2.3 and 2.4, we find that although construction costs (both labour and materials costs) are higher in Denmark than in most other European countries, the Danish construction industry has achieved one of the highest productivity levels (as measured by total construction output per worker). Similarly, in Sweden, although labour and materials costs are also high, the Swedish construction industry has achieved high productivity levels. Germany has the highest labour costs in our sample of countries, and average levels of construction productivity. Despite having the lowest total building costs, the French construction industry has lower productivity than the other mainland European countries' construction industry in our study. Despite having the lowest labour

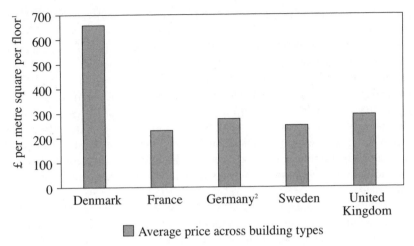

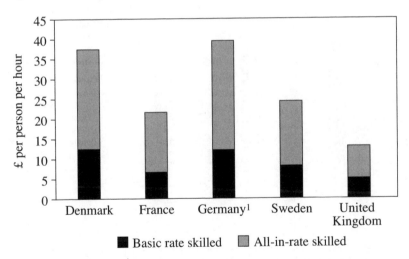

Notes:
1. Average price taken across the following types of building: heated office, air-conditioned office, factories/warehouses, high technology research, high rise apartments, shopping centres, provincial hotels.
2. Figures for Germany represent an average across Berlin and Frankfurt.

Source: Gardiner and Theobold (1998).

Figure 2.1   Comparative construction costs 1998

Note: 1.   Figures for Germany represent an average across Berlin and Frankfurt.

Source: Gardiner and Theobold (1998).

Figure 2.2   Comparative labour costs 1998

costs in our study, the UK construction industry also has the lowest productivity.

Our research findings suggest that the strength of inter-organizational cooperation may be responsible for the enhanced performance of the construction industry in some of the countries. The absence of formal R&D departments or formal research activities in many construction firms does not mean that innovation does not take place in the construction industry. However, because construction firms relate to many other industries in the supply stream, together with clients and with government through particular technology and information flows, construction industry innovation can only be understood in relation to the networks in which construction firms are embedded.

The findings in this chapter may lend support to the argument that firms operating under similar sector conditions (in terms of market structure, cyclical nature, entry and exit of firms) tend to adopt different strategic approaches to networks according to the nature of the different national institutional frameworks within which their production activities are

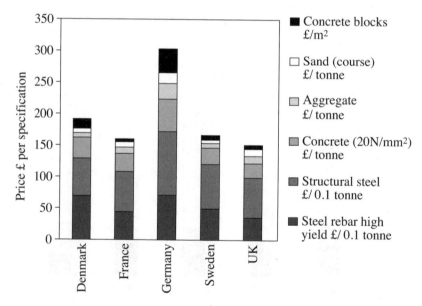

Notes:

1.   The figures exclude VAT and local taxes but include rates of delivery and local discounts.

2.   Costs for capital cities. Figures for Germany represent an average across Frankfurt and Berlin.

Source: Gardiner and Theobold (1998).

Figure 2.3   Comparative materials costs 1998

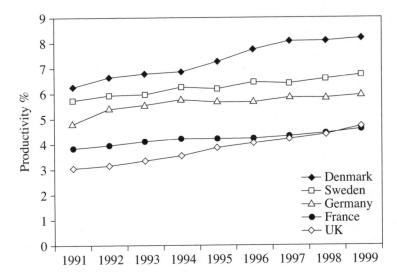

Note: Productivity expressed as a percentage of total construction turnover (in 1999 GDP) deflated by the number of employees in the industry.

Source: FIEC (1999), Eurostat (1998).

Figure 2.4   Comparative productivity, 1991–99

conducted. Also, stable long-term networks may be responsible for enhanced performance. This may be because strong linkages may encourage investment in relationship-specific assets, and may enable firms operating within a network to exploit production and exchange efficiencies not generally available to independent firms transacting on a short-term arm's length basis (Kester 1992).

This chapter has attempted to illuminate these mechanisms by reporting on the results of detailed case studies of inter-organizational relations supporting innovation in construction in five European countries. In countries where inter-organizational relations are strong, such as in Denmark and Sweden, the productivity of the construction industry is higher, despite high labour and materials costs. In Denmark, the government has taken an active role in promoting collaborations along the building supply stream through demonstration projects to encourage process innovation. Also, architects and engineers are actively integrated into the supply stream and have an important role in designing and incorporating new products. In Sweden, longer-term relations between construction firms and universities and with materials suppliers and manufacturers are responsible for process and product developments. At the other extreme, France and the UK, despite having some of the largest and most profitable construction firms, and some of the lowest total costs and labour costs respectively, have the lowest productivity levels. In France, while leading firms engage in international collaborations among contractors and with machinery producers to develop new technologies, all other inter-organizational relationships are weaker than in the other countries analysed. Similarly, in the UK, collaborations between contractors and clients are important but all other collaborations with other parties along the supply stream to improve the building process have not been strengthened. This may result in lower investments in innovation in the construction industry in the UK and France, thus leading to lower productivity and poorer performance.

These results suggest that construction firms make use of knowledge and technology that come from different organizations and combine them in ways that provide significant improvements in construction products or processes. More international comparative research is required to clarify the inter-organizational relations between the parties both on and off site and at different sectoral boundaries that contribute to innovation in construction. Also, given the importance of inter-organizational relations for innovation and competitiveness, government can take steps to create a supportive environment within which inter-organizational co-operation can develop most effectively.

NOTES

1. In Germany there are twice the number of architects than in the UK (one in every 1000 people is an architect). Prior to reunification, 60 schools of architecture taught 50 000 students each year despite a 10 per cent level of unemployment amongst architects (Building 1994). The qualification in Germany is more technical and rigorous than, for example, in the UK because when qualified, the architect is responsible for obtaining the building permit, designing the project and supervising construction. Management skill is within the architect's remit, reflected by the fact that 80 per cent of contract managers are qualified architects (Building 1994).
2. At present, subcontractors are employed on the lowest cost tender, but the Director of Business Development of Strabag believes that the client's attitude and procurement practices must change to extend the profit-sharing ideology further down the supply chain. Importantly, clients must no longer look at price alone. To test whether clients would be amenable to different forms of contract, Strabag sent a questionnaire to 100 of its clients. Only one third said that price is the primary reason for giving a contract and two-thirds argued that quality, goodwill and timescale are just as important.
3. Using the 'Laserblast Lama', the cleaning subsidiary of GTM made a turnover of FF1.5 million in the first year with one laser, and FF2.5m with two lasers in the second year. This technique is now being applied to industrial maintenance.
4. In fact, the concept of partnering – or at least the circumstances under which contracts could be awarded without lowest-cost tendering – was recognized previously by the Banwell Report (The Placing and Management of Contracts for Building and Civil Engineering Works) in 1967. If, for example, there was a good working relationship between the client and the contractor which had been developed over a period of time or where projects had been previously completed by the contractor on time at the desired quality for a reasonable price (Harvey and Ashworth 1997).

PART II

Adoption and diffusion of sustainable
technologies in construction

3. Sustainable technologies and the innovation–regulation paradox: the case of natural thermal insulation

INTRODUCTION

There has been substantial investment by governments and firms across Europe in the development of technologies and products that support sustainable building and sustainable urban regeneration. Despite a general slow rate of progress, there remain marked differences between individual countries, which suggests that there are sets of factors and institutions that inhibit or facilitate the adoption of sustainable technologies. Any attempt to promote environmentally-responsible house building and renovation must consider carefully the distribution of risk and decision-making power reinforced by the system of production, regulation, ownership and finance. This is important because the rate of adoption of new technology in construction determines not only the future competitiveness of the sector but also the strength of the economy's productive structure and affects the general level of employment and future skill requirement. A more sustainable and energy-efficient domestic sector is vital for addressing the problems of climate change. Compared to other European countries, progress in the UK toward increased social and environmental considerations in the construction industry has been relatively slow. Despite many government initiatives, there remain important institutional barriers (such as the corporate governance structure, profit motivation and extent of shareholder ownership) particular to the UK, which hinder the support of sustainable technologies by private-sector firms. This chapter assesses the case of the UK, which is of particular interest because a change in the attitude of the private sector combined with international, national and local government initiatives during the 1990s has prompted a more sustainable agenda in construction. The years since 1990 have also seen many technological innovations in energy efficiency. Despite the paradox of innovation and regulation (since the former is concerned with re-writing the rules and replacing the incumbent products and processes specified by the latter), both innovation and regulation are required to move the industry toward a more sustainable future.

Due to the fragmented structure and project-based nature of the construction industry, the effective adoption of innovation, and particularly of environmental innovation, requires the participation and collaboration of all parties in the industry. Sustainable innovation in the construction industry can be defined as those products and processes that either reduce the energy requirements of buildings and/or reduce the environmental impact (the so-called 'environmental footprint') of buildings, and structures. Product innovation would include, for instance, the use of natural materials, recycled/renewable materials or low-embodied-energy materials. Process innovation would include resource-efficient construction methods such as the minimization of energy and waste, maximization of recycling, local sourcing of materials and the use of brownfield sites. Sustainable innovation also includes innovative design, for example, designing a building to maximize passive solar gain. Sustainable technologies must also have a social and economic dimension. The social dimension can be in terms of intra-generational equity, improving the standard of living of the poorer sectors of society by for example, reducing the energy bills of social housing tenants. The economic dimension can be in terms of not compromising the need for private firms to maintain certain levels of profit, particularly in a low-profit margin industry like construction. In the construction industry, most sustainable product innovations stem from upstream product manufacturers and suppliers of building materials but all parties in the building chain have certain responsibilities to promote their adoption and use. It is the responsibility of the client to specify the use of technologies that reduce the consumption of resources over the lifetime of a building and to consider life-cycle costs in addition to capital costs. It is the responsibility of the engineer and the architect to interpret the client's requirements to include technologies that improve the design of the project. And it is the responsibility of the contractor to include technologies that improve the buildability of the project. For example, these improvements can be sustainable, involving a clean and efficient production process, use of low-embodied-energy materials and/or waste minimization. Implementation of sustainable technologies has been hindered in the past by a 'vicious circle of blame' whereby each actor in the industry blames each other for not building environmentally-friendly buildings (Cadman 1999).[1]

Although most European countries have seen a move toward more sustainable building, the UK provides a good example of public and private stakeholders working together to introduce wide-reaching reform in the way the construction industry operates. In addition to government initiatives, important drivers behind this change have included pressure from non-governmental organizations (for example, the Forum for the Future) and the changing attitude of leading firms and of the City to environmental performance indicators. In recent years, contractors such as Carillion and Morrison have published environmental reports with their annual reports, reporting their 'green' credentials and comparing their environmental performance over consecutive years (for example, see Carillion 2000). The corporate governance structure of British industry, characterized by its emphasis on delivering profits and dividends to shareholders, has hindered prioritization of environmental concerns because of the apparent trade-off between economic and environmental bottom lines. The mid-1990s, however, saw evidence to suggest that the relationship between environmental and financial performance need not be in conflict. According to Edwards (1998), in the building materials and merchants sector of the *Financial Times* All Share index, green building firms performed better than non-green firms between 1992 and 1996. During the late 1990s, two important drivers have emerged to redress this economic imbalance further. First and foremost, the City and shareholders began to express interest in sustainable issues (for example, Cowe and Williams (2000) note that three-quarters of City investors say that the City is taking ethical and green issues more seriously).[2] Second, firms felt the need to be (seen to be) environmentally conscientious to secure future contracts (for example, in the near future it is likely that firms may be required to demonstrate their environmental credentials in order to secure a place on public sector tender lists from government departments and agencies).

The state, both at the national and local authority level, is the single most influential party in supporting the achievement of sustainability targets through its position as the largest client of the construction industry, its capacity to offer fiscal incentives and ability to 'move the goalposts' by undertaking a review of building regulations. Also, it has an important role to play as principal educator and disseminator of information to the industry or as market leader, with the ability to prototype innovative solutions through demonstration projects. The construction industry is influenced by technical regulations governing products and processes, planning and environmental regulations governing the finished product, and health and safety regulations governing the welfare of workers during the construction process (Gann 1999). It has been argued that more building regulations means that houses are built uniformly and firms compete on price alone, leading to increased risk in the use of new products and processes (Blackley and Shepard 1995, Pries and Janszen 1995, Tatum, 1987). Private firms will naturally oppose increased environmental regulation since the direct costs are clear whilst the potential future savings are unknown (Wubben 1999). However, regulations are in place to protect the interests of the public and the environment, to maintain minimum quality standards, to provide a level playing field for firms to compete and to provide a buffer for innovative firms until new technologies are proven and economies of learning reduce their costs (Porter and van der Linde 1995). Whilst conventional wisdom tells us that environmental regulations impose significant costs, are responsible for slow productivity growth and hinder firm performance, recent revised opinion has supported a net positive impact of environmental regulation (Jaffe et al. 1995).[3]

Although, in general, there is no empirical analysis that offers convincing evidence to support the assertion that environmental regulation stimulates innovation across the board, the building industry offers good examples of increased resource productivity and lower finished product total cost in the presence of stricter environmental regulation (Jaffe et al. 1995, Welford and Starkey 1996). For example, in Sweden and Germany, where there is considerably stricter environmental regulation, total building costs are below those in the UK, despite higher material costs and labour costs (see Figure 1.1 in Chapter 1). In these countries, construction processes have been improved to outweigh the component costs of building. Nevertheless, regardless of whether environmental regulations help or hinder innovation in industry, they affect the competitive behaviour of firms and the competitive dynamics of the industry imposing new costs, investment demands and opportunities to increase production and energy efficiency (Shrivastava 1995).

This chapter examines the importance of regulation and innovation in reducing the energy consumption of domestic buildings. The case of an energy-saving technology – natural thermal insulation materials for cavity wall insulation – suitable for widespread use in residential buildings is assessed. This technology was selected since it may be expected to make a significant contribution to sustainable building and regeneration on its own account and because it has the potential to demonstrate at a more general level the underlying obstacles and facilitative factors which influence the innovation process. Where appropriate, international benchmarks and the experience of other European countries will be considered. Thermal insulation is one of only a number of options that could be employed to increase domestic energy efficiency (other include improving the performance of the heating and hot water systems or the efficiency of boilers and lighting), though it is certainly the most cost-effective way of reducing energy consumption and carbon dioxide emissions. Through reducing energy consumption by increasing the level of insulation in existing buildings and installing higher thermal values of insulation in new build, non-renewable fuel supplies can be conserved, reducing the amount of pollutants created in the burning of fossil fuels (for example, carbon dioxide, nitrogen oxides and sulphur dioxide). The chapter is organized as follows. The first section describes the importance of the domestic sector in combating climate change and reviews the action undertaken by the government at the national and local level in the UK. The second section explores innovation in the thermal insulation industry. The conclusion looks at factors inhibiting and facilitating the use of new sustainable thermal insulation and draws policy implications from the analysis.

CLIMATE CHANGE, THE UK DOMESTIC SECTOR AND THE ROLE OF THE GOVERNMENT

There has never been a more important time to understand the innovation process of sustainable technologies and encourage the implementation of energy-efficient technologies in housing. The world is undergoing significant climate change and global warming, due to increased levels of greenhouse gases in the atmosphere raising the temperature of the earth above its natural equilibrium level. Carbon dioxide is the single largest contributor to greenhouse gases; other important greenhouse gases are methane, nitrous oxide, hydroflourocarbons, perfluorocarbons and sulphur hexaflouride. Although emission levels of these gases are significantly lower than carbon dioxide, they exert a much larger contribution per unit gas.

During March 2000, the draft UK Climate Change Programme outlined a series of policy measures, with an emphasis on carbon dioxide emission reduction (including regulation, economic instruments, education and expenditure), to cut greenhouse emission by 21.5 per cent by 2010 (DETR 2000a). Current estimations suggest that by 2010 UK emissions of the six greenhouse gases will be 13.4 per cent below 1990 levels. However, carbon dioxide emissions are forecast to fall by only 7 per cent, with levels rising after 2000 with the closure of nuclear power plants and increasing economic growth.

According to the Building Research Establishment (BRE), in 1996 the energy use of buildings (for heating, lighting and cooling) accounted for 50 per cent of the UK's primary energy consumption, equating to 45 per cent of total UK carbon dioxide emissions, around 25 per cent of sulphur dioxide and nitrous oxide emissions and 10 per cent of methane emissions (BRE 1999). The UK domestic sector is responsible for approximately one quarter of total carbon dioxide emissions (see Figure 3.1).

The use of direct economic instruments to increase fuel bills is deemed to be politically unacceptable and the UK government is using a two-prong policy of education and stick-and-carrot fiscal measures. First, through programmes such as the Energy Efficiency Best Practice Programme and public information campaigns such as the 'Are you doing your bit?' campaign, firms, organizations and households can be educated and informed about the costs and benefits of energy-efficiency alternatives. Second, by promoting the installation of energy-efficient measures using financial incentives and, where

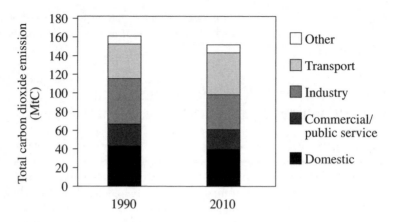

Source: DETR (2000a).

Figure 3.1   Carbon dioxide emissions by sector, 1990 and 2010

necessary, regulation, energy use can be decreased. Following the Earth Summit in 1992, UK local authorities responded to the Agenda 21 sustainable development commitment by implementing sustainable energy strategies within their region. Further evidence of a sustainable agenda came with the Home Energy Conservation Act (HECA) and the introduction of the Building Regulations for the Conservation of Fuel and Power, requiring new and renovated houses to achieve minimum Standard Assessment Procedure (SAP) rating. Financial incentives are being offered, for example, through Home Improvement Agencies and the Home Energy Efficiency Scheme (HEES). Further, in April 2000, the UK government announced a strategy for more sustainable construction (a collaborative framework between government and industry, which identified action areas and suggested performance indicators) that will complement the policies outlined in the Climate Change Programme.[4] These include proposals for fiscal measures (for example, the landfill tax), changes to public sector procurement, development of the construction industry's image, waste minimization and resource conservation.

Notwithstanding fiscal incentives and other government initiatives, minimum energy efficiency regulations have largely determined the extent and type of thermal insulation in the UK and all the rest of Europe. In the UK, the regulations governing thermal insulation standards are included in Part L of the Building Regulations devoted to the conservation of fuel and power in buildings. Planning policy guidance and building regulations have been recently overhauled to reflect the aims of the Climate Change Programme. Building regulations requiring energy conservation in domestic regulations were introduced in 1965 and amended in 1976, 1982 and 1990. A 1998 consultation paper for the DETR suggested higher thermal insulation standards for new properties and in the refurbishment and operation of existing buildings (Oscar Faber 1998).[5]

In July 2000, two consultation papers on the conservation of fuel and power in the English and Welsh and Scottish building regulations were published (DETR 2000c, Scottish Executive 2000). Regulatory changes were being considered for phased implementation in the UK beginning in late-2001.[6] Alongside the replacement of the SAP by the Carbon Performance Index (CPI) for domestic buildings, the most significant proposals included increases in elemental and target U-values and efficiency and control improvements in heating and lighting.[7] The requirements extended the definition of material alteration to include more retro-fit work within the scope of the regulations. Trade-offs between efficient boilers and fabric insulation were also being considered: for example, exposed wall U-values of just $0.30W/m^2K$ will be required if a SEDBUK (Seasonal Efficiency of a Domestic Boiler Database) boiler is used. This is part of a trade-off package, where the poorest acceptable U-values will fall to $0.7W/m^2K$ if compensated by the performance of other elements. These proposed regulation changes to domestic dwellings are estimated to reduce carbon emissions by 1.32 MtC by 2010, over half of which will stem from alterations to existing dwellings.[8] Figure 3.2 shows the existing and new regulations (to be implemented by 2008) governing minimum insulation levels in the UK. It also shows the minimum regulation standards across countries in Northern Europe. Most of the countries shown have tightened their minimum standards recently, in response to climate change or energy efficiency commitments. For example, in Denmark new building codes were introduced in 1995 to cut space heating demand by 25 per cent (Kerr and Allen 2001). Also in Germany in 1995 the federal government reviewed thermal insulation requirements to 'limit carbon dioxide emissions by the more efficient use of energy' (Institute of Building Control 1998).

SUSTAINABLE THERMAL INSULATION TECHNOLOGIES

Developments in materials technologies present a technological solution to improved thermal insulation. There are a number of barriers that prevent the uptake of new technologies, some of which are particularly pertinent when considering sustainable technologies. In addition to the increased risk and the lack of information for industry and the wider public of new technologies, the costs involved in using a new technology are the single most important barrier. Deregulation of the UK gas and electricity industries may have sent out a wrong signal about energy conservation, removing the financial constraint on wasting energy. Arguably a bigger problem is that promoters or those financing building projects give more consideration to the capital cost of thermal insulation as opposed to its life-cycle cost or environmental cost. To

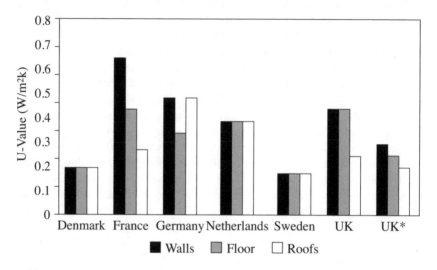

Notes:
1. UK* shows the standard elemental U-values existing before the implementation of the new building regulations governing energy conservation in 2003. The figures are based on a SEDBUK boiler being present in the dwelling. By 2008, the suggested insulation values are 0.25 W/m²K, 0.22 W/m²K and 0.16 W/m²K for walls, floors and roofs respectively.
2. Figures for 1995 for Denmark, the Netherlands and Sweden. 1997 figures for France and Germany.
3. In Sweden, a formula is used to calculate the maximum thermal resistance: $0.18 + 0.95A_f/A_{om}$ represents the maximum average thermal resistance where A_f is the aggregate area of the windows, doors etc, and A_{om} is the aggregate area of the enclosing elements of the structure in contact with the heated indoor air. If we assume the following dimensions of a detached house:
   $A_f = 30$
   $A_{om} = 180$ then the maximum average U-value across the building would be 0.34.
4. In France, another formula is used to calculate the thermal insulation levels. Depending on the type of heating (electricity or otherwise) and climatic zones (H1, H2 & H3), the area of roof, floor, walls, doors and windows are multiplied by constants varying between 0.25 and 3.5. For example, the surface area of the walls is multiplied by a constant 0.6 if the house is heated by electricity, which varies between 0.65 and 0.8 if heated by other sources according to the climatic zone.
5. UK figures assume a SAP rating of over 60.

Sources: Data for all countries from the Institute of Building Control (1996-1998), proposed 2003 UK data from the DETR (2000c).

Figure 3.2   Comparative regulations governing thermal insulation standards of exposed elements in six European countries

calculate the life-cycle cost of a material, capital costs must be considered alongside maintenance costs, the materials' availability, installation costs and forecast lifespan. One can also calculate the cost of the material in terms of its triple bottom line, ensuring that environmental and social considerations are considered in conjunction with the pure economic cost (including the externalities generated in both the production and use of the materials and considering the liability and risk issues involved with the safety of those who build, use or occupy the building). An analysis of a number of conventional and sustainable technologies currently available, including an examination of their performance, cost and energy efficiency potential considering the environmental impact of their production, installation and use follows.

Since thermal insulation is a hidden innovation (in as much as it has no aesthetic properties), it is a functional technology, the adoption of which depends on its performance and price. The performance of insulation materials depends primarily upon their ability to trap still air and although cavities and surface resistances are important, the thermal resistance of construction materials is the most significant factor. Thermal conductivity, or K-value (the reciprocal of the thermal resistance), measures heat flow through a given amount of material (for example, a good insulator will have a low K-value).[9] There are three different forms of insulation used in the control of heat flow: reflective, resistive and capacitive insulation.[10] Resistive insulation materials are the most common and are produced in two types, fibrous materials and foams. Fibrous materials include mineral wool, glass fibre batts and quilts, and organic fibres such as cellulose. Foams include expanded polystyrene (EPS), extruded polystyrene (XPS), polyurethane and urea formaldehyde. Foams are available as rigid or semi-rigid slabs and can formed in situ (that is, injected into cavities). In 1992, a study conducted by the Building Services Research and Information Association (BSRIA) and the BRE found that the materials principally used in cavity wall insulation in non-timber houses were glass mineral wool slab, rock mineral wool slabs and extruded polystyrene slab, accounting for 85 per cent of market share; for external wall insulation, EPS foam and mineral fibre account for 85 per cent of total market share (Bell College of Technology 1994). The thermal conductivity of these products, relative to other materials, is shown in Table 3.1.

The widespread use of these materials can be largely explained by their low K-values, efficiency and relatively low cost, encouraged by the construction industry's preference for tried and tested materials, the performance of which has been monitored and proven over many years. However, as Table 3.1 highlights, these materials fit uncomfortably alongside the concept of sustainability, producing a significant environmental impact during their production and use. For example, rock wool and glass wool are produced by a similar process involving the combination of raw materials through intense

Table 3.1 Thermal conductivity of insulation materials

	Thermal Conductivity (W/mK) at 10°C	Density (kg/m³)	Environmental Impact		
			Production	Use	Total
Chemical insulation material products					
Phenolic foam	0.022	60	5	3/4	5 3/4
Rigid polyurethane foam	0.023	35–50	5	3/4	5 3/4
Extruded polystyrene foam	0.026	28–45	5	1/4	5 1/4
PVC foam	0.029–0.048	40–300	5	1	6
Glass mineral wool	0.031–0.037	16–80	3 1/2	1	4 1/2
Rock mineral wool	0.033–0.037	23–80	3 1/2	1	4 1/2
Expanded polystyrene	0.033–0.038	15–30	5	1/4	5 1/4
Natural insulation material products					
Vital	0.034	40	1/4	0	1/4
CR flax	0.037	30	1/4	1/4	1/2
CR wool	0.037	16	1/4	1/4	1/2
Wool	0.037	–	1/4	0	1/4
Cellulose fibres	0.037	–	1/4	0	1/4
Cork	0.038	112	1/4	1/4	1/2
Homatherm	0.040	85	1/4	0	1/4
Isoflac	0.040	40–70	1/4	0	1/4

Gutex Thermosafe	0.040	160	$\frac{1}{4}$	0	$\frac{1}{4}$
Gutex Thermowall	0.040	160	$\frac{1}{4}$	0	$\frac{1}{4}$
EMFA coconut fibre boards	0.045	124	$\frac{1}{4}$	0	$\frac{1}{4}$
Gutex Happy Step	0.050	260	$\frac{1}{4}$	0	$\frac{1}{4}$
Exfoliated vermiculite	0.066	109	$1\frac{3}{4}$	0	$1\frac{3}{4}$

Note: Environmental impact, both in terms of production and use, ranked 1–5 where 5 represents the most environmentally damaging.

Source: Thermal Insulation Manufacturers and Suppliers Association (2000), Construction Resources (2000) and Woolley et al. (1997)

heat. Mining is required to extract the raw materials and the production process is energy-intensive, creating emissions of fluorides, chlorides and particulates and releasing solvents and volatile organic compounds such as phenol and formaldehyde (Woolley et al. 1997). In addition, sulphur oxides and nitrogen oxides are produced contributing to acid rain and causing photochemical oxidants (Curwell and Mach 1986). In terms of its use, fibreglass has been measured above some landfill sites and there is concern that it may be an atmospheric pollutant because of its non-biodegradable properties (Curwell et al. 1990). There is also inconclusive evidence surrounding the carcinogenic properties of glass fibre since it contains small amounts of the harmful small-sized fibres found in asbestos in addition to oil and resin binders which limit their harmful release (Curwell and Mach 1986, Curwell et al. 1990, Woolley et al. 1997).

Plastic foams are even more environmentally damaging than their fibrous alternatives, particularly in terms of their global warming potential, partly because of the high-embodied energy raw materials and partly because of the use of blowing agents. The raw materials, oil and natural gas, are non-renewable resources and their use, associated with emissions of oils, phenols, heavy metals and scrubber effluents, account for over half of all toxic emissions into the environment (Woolley et al. 1997). For example, EPS is created by fusing polystyrene with pentane, XPS by combining polystyrene with blowing agents. Blowing agents are used to increase the energy efficiency of the material, expanding the polymer matrix and adding to the thermal conductivity through the blowing agents inherent K-value but impose a heavy cost on the environment.[11]

As shown in Table 3.1, there is no shortage of alternative materials, many of which are marketed as 'sustainable' or 'natural' alternatives. In terms of the environmental impact during production and use, the table also illustrates the contrast between the incumbent technology and the natural alternatives. All natural insulation materials are produced from renewable plant or animal resources, have low embodied energy, use only natural additives such as borax (sodium tetraborate) or potato starch (which means that there are no toxic by-products during their manufacture and no health problems during installation), and are fully biodegradable (that is, they contain no toxic or synthetic chemicals) (Construction Resources 2000). For example, 'Vital' is cellulose insulation in batt form made from oxygen bleached wood pulp and viscose fibres. It is bonded with a food-grade cellulose-based binder and treated with pH neutral boron liquid to protect against fire and decay. The material is able to absorb up to 20 per cent of its weight in moisture and is non-toxic and free of emissions. The production process of Vital produces 40 per cent more energy than it consumes and it can be recycled or biodegraded at the end of its life. In addition, the installation is free of health risks, the insulation can be handled, does not scratch or cause itching and does not require a dust mask to be worn. Cellulose fibre insulation is made from processed waste paper and treated with borax (sodium tetraborate) to guard against fire and insects (Harland 1993). The insulation can be installed by hand or sprayed and is commonly used in 'breathing wall' timber frame construction and in lofts (Woolley et al. 1997). The production of cellulose fibre insulation does not cause any pollution and has a relatively low embodied energy (the total amount of energy used in the raw materials and manufacture of a certain quantity of material). In fact, the only negative environmental impact stems from the energy used in the materials production (Curwell and Mach 1986).

Table 3.1 also shows, however, that the natural thermal insulation materials are poorer performers in terms of their thermal conductivity. There are two contrasting issues that must be considered to evaluate the true sustainability of the thermal insulation technologies. One issue concerns the direct environmental impact of the production and use of thermal insulation materials. The above analysis has shown that many conventional materials have high embodied energy and have properties that affect health and prevent the materials biodegrading or being re-used. There is significant environmental damage imposed by the production (for example, mining of raw materials, energy intensive production processes in fibrous insulation, use of HFCs and HCFCs in foam insulation) and use (for instance, materials that are non-biodegradable or have carcinogenic properties) of these materials. Some natural insulation materials offer an alternative with significantly fewer negative externalities (for example, no mining of raw materials, no consumption of limited resources, no health problems during or following installation and no synthetic ingredients preventing biodegradation). The range of natural insulation products available today demonstrates that there is no lack of innovation in thermal insulation materials.

However, in achieving sustainability targets through increasing the energy efficiency of the domestic sector, the most important consideration is the need to reduce energy consumption and this depends on the thermal insulation material's performance in limiting heat loss. There is evidence to suggest that the energy savings in terms of a natural insulation material's embodied energy does not offset energy savings over a conventional material's lifetime performance (Heath 1999). It has been estimated that despite the higher embodied energy and higher capital cost of conventional materials such as plastic foams, their far superior insulation performance results in positive economic, environmental and social (in terms of lower fuel bills) benefits when compared to their fibrous alternatives (Heath 1999).[12]

Therefore, natural insulation materials currently do not offer a good enough performance to be considered a credible alternative to the incumbent technology. Nor will they contribute toward more sustainable buildings.

Indeed, the only way to reduce significantly the energy consumption of domestic buildings is to increase the minimum insulation levels. However, as discussed above, the private costs and social benefits mean that the building industry has no incentive to build above the minimum standards. The experience of other countries suggests that sustainability in construction requires the tightening of UK regulations. The process is particularly effective when the new regulations are announced with a sufficient time lag and the process is staggered. This provides an important stimulus for innovation since the thickness of thermal insulation material cannot be simply increased because of space considerations (and even if this was the case, innovation would then be needed in the design stage of the construction stage).

IMPLICATIONS

The regulatory changes will help achieve the Government's CO_2 target, improve housing energy efficiency and contribute towards sustainable construction and managing the effects of global warming. It is estimated that 1.32MtC emissions can be saved by 2010 as a result of the new regulations and that new housing built to the new standards will contribute between 25 per cent to over 30 per cent less carbon emissions (Building 2000, DETR 2000c, Harper 2000). However, because the number of new-build homes only increases the housing stock by 1 per cent each year, the reduction as a proportion of the total housing stock is in fact very small. Significant improvement in the energy efficiency of the housing stock will be needed since houses built as recently as the end of the 1980s need a decrease of over 50 per cent heat loss to meet the new standards.[13] Housing associations have been encouraged to conduct energy efficient and environmentally sound refurbishment of their existing stock but a more proactive stance on behalf of home owners is needed and is likely to require substantial subsidies from government and local authorities or industry.[14] The cumulative effect of more stringent regulations applied to new-build housing and improvements to the existing housing stock will gain momentum up to 2010 and will accelerate thereafter.

It could be argued that the proposed regulatory changes do not go far enough and that the only way to meet sustainability and innovation targets is to account for the environmental cost of carbon dioxide emissions through building regulations, including increasing them toward their Scandinavian equivalents. This would be particularly effective in the colder northern climate of Scotland. Furthermore, one can argue that stringent prescriptive U-value minimum standards are required to make the industry build to a more sustainable standard. This research does not support the case for the implementation of U-value trade-offs in the UK (available, for example, where efficient gas-fired central heating systems are installed). Although trade-offs give designers more flexibility, they impose another level of complexity and detract from the importance of imposing lower minimum U-values. With a prescribed minimum thermal insulation level, the building's energy efficiency over its lifetime can be calculated and guaranteed and will not be compromised by the replacement of energy efficient components included initially to benefit from the trade-offs. In the presence of trade-offs, continued energy efficiency could be monitored by an extension of the proposed 'MOT test', which the DETR plans to conduct for larger buildings, or by requiring an energy assessment to be included in the information sellers must provide when marketing their houses. Because of the energy bill saving implications of these proposals, government must work closely with industry, housing associations and housing authorities to ensure the efficient retro-fit (including the modification, renewal and extension) of existing buildings.

This chapter has examined the paradox between innovation and regulation and its implication for the adoption of sustainable technologies in the domestic sector of the construction industry. Using thermal insulation as an example, the chapter has examined the underlying innovation process of sustainable technologies and outlined the principal factors inhibiting and facilitating their adoption, highlighting the fact that both innovation and regulation are needed to promote a more sustainable future for the construction industry. Though the UK is used as an example, the conclusions are applicable at a general level, particularly those concerning the levels and type of regulation, the need to evaluate the environmental costs of innovative materials more thoroughly and effectively and the need to engage all actors in the construction industry through education and the dissemination of good practice. Innovation alone will not succeed in countering the problems of energy inefficient buildings. Notwithstanding fiscal and other incentives, tighter thermal insulation regulations need to be applied to both new and, more importantly, existing buildings to reduce energy consumption and contribute towards a more sustainable domestic sector.

NOTES

1. Cadman (1999) explains that contractors argue they could provide environmentally efficient buildings but complain that the developers do not specify them. Developers argue they would like to specify more environmentally efficient buildings but investors will not pay for them. Investors argue that they will not pay for these because there is no demand from client occupiers to justify them.
2. Ethical investments in the UK have increased throughout the 1990s and are currently valued at over £3.3 billion. Although this only represents 1 per cent of the market, the figure is likely to rise with 2001 UK legislation on pension fund disclosure (see Cowe and Williams 2000, MacGillivray 2000).

3. Porter and van der Linde (1995) argue that 'properly designed' environmental regulation can stimulate innovation, lowering product cost or improving product value, allowing firms to be more productive (for example, in terms of raw materials, labour and energy) offsetting the costs of reducing the environmental impact. Though this does not equate to technology-forcing, it is technology-facilitating.

4. The industry is to add another KPI (Key Performance Indicator) on project sustainability, measuring waste, energy, water, ecology, transport and recycling (DETR 2000b).

5. Recommendations for new buildings included significantly lower K-values for walls and windows (a good insulator has a low K-value), minimum efficiency standards of ventilation systems and minimum number of compact fluorescent light bulbs. Recommendations for existing buildings included new building standards that could be applied during any major refurbishment or change of owner/tenant. Further suggestions have included periodic (every 7–10 years) energy surveys of homes that belong to the same owners for long periods of time and 'MOT tests' for existing buildings, to check if the building fabric and internal systems operate as intended (ENDS 1998, Oscar Faber 1998). These checks would be implemented in conjunction with a system of sanctions for inefficient operators.

6. Under a two-stage programme, the building regulations governing U-values would tighten: for example, exposed wall U-values would tighten from $0.45 \text{ W/m}^2\text{K}$ to $0.3 \text{ W/m}^2\text{K}$ by 2004. Similar changes apply to the ground floor, roof, exposed floor, windows, doors and rooflights.

7. The CPI replaced the SAP Energy Rating Method giving designers and builders more flexibility in meeting annual carbon targets. The SAP rating will still have to be calculated though no minimum standard will be required.

8. Although the principal benefits are seen in terms of meeting the Government's carbon emission targets, more direct benefits will be available to the household through reduced energy bills. In existing buildings, a 25 per cent saving in energy equates to £125 per year for a typical domestic dwelling, representing a carbon saving of 0.15 tonnes per year, depending on people's choice between lower fuel bills and increased warmth. A reduction in carbon emissions by more energy-efficient new-build housing by 2010 has been estimated at 0.25MtC per year in England and Wales and 0.065MtC per year in Scotland. As for the cost impact of the proposed changes, following the two-stage introduction of the regulations, it is estimated that the price of a new detached and semi-detached house will rise by between £900 and £1400, smaller types of houses increasing in price by between £600 and £1100. It must be noted though, that any predictions of future costs are inherently inflated since they fail to account for improved technology. Data predictions from the DETR (2000c) and Scottish Executive (2000).

9. An insulation material's weight, strength to weight ratio, convective heat loss, settling and loss of insulating capacity, thermal and vapour resistivity, water absorption properties and resistance to moisture transmission and fire credentials are also important. For example, the weight of an insulation material is important since, for example, sagging can occur in ceilings. Convective heat loss in insulation caused by air currents is rare but can occur when different temperature air currents below and above the insulation cause small 'convection loops' within the insulation. Standardization of fire regulations is presently being undertaken and the differences that currently exist between incumbent regulations influence the choice of insulation materials. The core insulating material can affect its fire performance; so too can the choice of blowing agent. Additional fire retardants may be necessary and these will add to the cost of the insulation. For further details see EREC, (1995a) and Caleb Management Services (1997).

10. Reflective insulation can be used when the dominant heat transfer is by radiation. Radiant barrier installations have been used since the 1930s in the USA as an inexpensive way of protecting buildings from undesirable heat gain (EREC 1995b). Radiant barriers use reflective foil, for example, aluminium foil (which has low absorptance and low emittance) to block radiant heat transfer. Capacitive insulation is distinguished from resistive insulation because, rather than providing an instantaneous effect, it affects the timing of heat flow.

11. Clorofluorocarbons (CFCs), constituting any various gaseous compound of carbon, hydrogen, chlorine and fluorine, have been used as the preferred refrigerant since the 1930s

and, in more recent years, the preferred blowing agent. CFC-11 has a thermal conductivity of 0.017 at 10 per cent and can improve the thermal performance of XPS. The significant ozone depleting properties and global warming potential of CFCs was highlighted in the 1980s and, following the Montreal Protocol in 1987, CFCs were phased out to be replaced by hydrofluorocarbons (HFCs). CFCs have a Global Warming Potential (GWP) 4000 times that of carbon dioxide. HFCs, such as HFC-245fa, have a GWP 820 times greater than carbon dioxide and HCFC-141b has a GWP 630 times greater. In addition, HCFCs also have significantly lower ozone depleting potential, 0.11 compared to 1 of CFC-11. Also, there are particular hazards in the use of polystyrene foams in terms of the emissions of carbon monoxide, carbon dioxide, smoke and water vapour if the material is exposed to fire. For a more in-depth examination of the principal thermal insulation materials used in the UK see Bell College of Technology, (1994). For more information on the use of blowing agents see Caleb Management Services, (1997).

12. For instance, Heath (1999) provides quantitative evidence of the superiority of phenolic foam in economic, environmental and social terms. He compared the costs and benefits of phenolic foam insulation and rock mineral wool fibre against no insulation for pitched roof insulation. As Table 3.1 shows, rock mineral wool fibre has similar performance properties to the best natural insulations. With a benchmark of no insulation, phenolic foam delivers net energy savings of 485000 kWh over a 50-year period compared to 446000 kWh for rock mineral fibre. Heath calculates this to equal a saving of 139 and 127 tonnes of carbon dioxide equivalent respectively over the 50-year lifetime. The energy savings in use significantly outweigh the embodied energy costs associated with their production, calculated by Heath to be 6100 kWh and 2200 kWh respectively. Even more conservative calculations estimate that the ratio of thermal insulation energy saving to energy investment is 12 to 1 per year. In financial terms, over a 50-year period, phenolic foam offers savings of £10 394 compared to rock mineral fibre with £9321. Therefore, although phenolic foam has a higher capital cost (£653 as opposed to £356), it is cheaper in the long-term.

13. Harper (2000) estimated that 99 per cent of homes will be under-insulated according to the new regulations at the end of 2000.

14. For example, the National Home Energy Rating estimate that a 1930s detached house without cavity wall insulation costs £1000 more per year to heat than houses built to current existing standards. Over a two-year period, £10 million of grants from the government industry, regional electricity firms and local authorities provided assistance to occupiers of properties with inadequately insulated cavity filled walls (Allder 1999).

4. Factors enabling and inhibiting sustainable technologies in construction: the case of active solar heating systems

INTRODUCTION

Greenhouse gases (GHGs) represent the most significant anthropogenic influence on climate change. Across Europe, the domestic sector is one of the largest users of energy, accounting for over a quarter of final energy consumption (EEA 2001a). Domestic sector GHGs are predominantly attributable to the energy required for space and water heating – across the EU, 84 per cent of household energy consumption stems from space and water heating (EEA 2001b). Improvements in the energy efficiency of housing and electrical appliances has meant that in Northern European countries such as Denmark and the Netherlands in particular, but also in France and Sweden, energy consumption per dwelling has fallen since the mid-1980s. In others, such as the UK and Germany, energy consumption has risen slightly: 1 per cent and 4.5 per cent respectively (see Figure 4.1). Overall, despite increases in energy efficiency, higher GHG emissions from the domestic sector can be explained by an increasing trend in the number of households and the average size of dwellings, coupled with a reduction in the average number of persons per household and falling domestic electricity prices.[1]

Mitigation strategies for greenhouse gas emissions have focused on improving the energy efficiency of buildings, both in terms of electricity use and space heating. As Chapter 3 has shown, the external temperature and the level of thermal insulation primarily govern the heating requirement of buildings and most European countries have tightened their building regulations during the 1990s. In addition to improving the thermal properties of the building envelope and developing mechanisms to encourage energy conservation, the use of new energy technologies in new-build and retro-fit residential buildings has the capacity to reduce significantly energy consumption. While in Chapter 3 we examined thermal insulation, here we examine active solar heating (ASH) systems for water heating, another

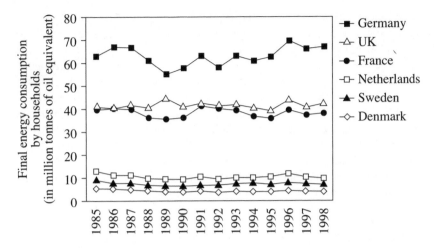

Source: EEA (2001b).

Figure 4.1   Final energy consumption by households across European countries between 1985–98 in million tonnes of oil equivalent

technology suitable for widespread use across new and existing buildings in the housing stock, which has the potential to make a significant contribution to sustainable building and regeneration. Their generally slow adoption can be attributed to high capital cost and unknown cost effectiveness, but these factors do not adequately explain variations in their use across Europe. This suggests that there are sets of more important factors and institutions inhibiting or facilitating their adoption.

The ASH industry grew from the oil crises of the 1970s fuelled by government sponsored research and development. The slump of energy prices in the 1980s reduced demand and most new solar firms died out. Surviving firms remained in business through improving their products and production processes (for example, in terms of quality and efficiency). Arguably the most significant innovation over this period was the evacuated tube collectors, which increased the efficiency of the ASH systems. Encouraged by the adoption of climate change policies across Europe and a more general environmental focus from government, industry and the public, the application of solar thermal technologies increased during the 1990s. Yet whilst countries such as Germany and Denmark have increased their total installed solar collector surface area year on year since the mid-1990s, the cumulative surface area in the UK and France has consistently fallen. This chapter examines the structural and institutional factors behind these differentials and draws implications for the management of innovation by construction firms and government policy for those countries under-exploiting the potential of ASH systems. For that purpose, this chapter is structured as follows. The next section examines the general barriers to sustainable technologies in the construction industry and looks at the attempts made by European governments to promote energy efficient products and processes in the face of these barriers. The section on technological change and the mitigation of greenhouse gases examines active solar heating technologies in more detail and assesses the structural and institutional factors behind the different rates of adoption of solar technologies across Europe. The final section draws policy implications for the UK and Scotland in particular, though the recommendations are applicable to other European countries, such as France, currently under-exploiting the potential of ASH systems.

SUSTAINABLE INNOVATION AND THE ROLE OF GOVERNMENT

As argued in Chapter 3, sustainable innovation in the construction industry can be defined as those products and processes that either reduce the energy requirements of buildings and/or reduce the environmental impact of buildings and structures. There is a considerable number of factors that inhibit innovation in the construction industry, which tend to be exacerbated where sustainable technologies are concerned. These include, for example, the nature of the construction industry (fragmented, conservative, mature and with low profit margins), the characteristics of the final product (immobility, uniqueness, complexity and costliness) (Gann 1994, Nam and Tatum 1988), and the operating environment (highly regulated, high liability and litigation risk) (Blackley and Shepard 1995, Pries and Janszen 1995). Attempts to address some of these problems in the 1990s have focused on the promotion of alternative forms of procurement.[2] However, despite the advantages of closer inter-firm relations facilitated by the adoption of alternative contractual arrangements such as design and build there remain important barriers, the most significant of which are risk and cost.[3]

The risk premium associated with sustainable technologies predominantly stems from the low profit margins in the industry, the uniqueness and complexity of the final product and the regulated and litigious operating environment. Risk in the adoption of any new technology in the construction industry is based on using an un(satisfactorily)-tested product or process about which little is known and associated with the responsibility for the safety of those who build, use or occupy the building. Nevertheless, as discussed in the introduction, the high cost of sustainable technologies does not simply account for an insurance premium however. Higher unit costs are the inevitable result of limited production (for example, high development costs, absence of economies of scale, learning and experience, limited distribution outlets and high transport costs) and specialized distribution (for example, more difficult to source designers (engineers and architects), suppliers and subcontractors with the capability, experience and willingness to design, supply and install the new technologies).[4] The cost of sustainable technologies also reflects market imperfections, which do not take account of the environmental and social costs. But if we consider the triple bottom line, sustainable technologies have a net benefit: economic (for example, lower running costs and hence lower fuel bills), environmental (for example, reduced energy consumption and hence reduced greenhouse gas emissions) and social (for example, combating fuel poverty). However, in both the public and private sector, capital cost considerations are paramount in the face of profit margins and limited public resources. This means that there is a trade-off between ecology and the economy (see Porter and van der Linde 1995, Wubben 1999), with social benefits on one side and private costs on the other – costs which are borne not by industry but by the ultimate owner of the building (see Bordass 2000, Malin 2000). This type of argument is familiar to students of the environmental innovation literature, with private firms complaining that regulatory burdens increase costs and hinder the use of environmental technologies. Nevertheless, Jaffe et al. (1995) and Welford and Starkey (1996) provide support for the idea of the positive effect of environmental regulation in stimulating innovation and Cowe and Williams (2000) and in Chapter 3 we refer to the key role of the firm in promoting a new private sector driven sustainable agenda. Cowe and Williams refer to the firm's ability to take sustainable operational and strategic decisions (for example, reducing energy or transport costs, not conducting business with firms selling socially or environmentally unsustainable products) and in Chapter 3 we point to the changing attitude of leading firms, the City and government agencies to environmental performance indicators. Thus, arguably the challenge for policy makers is to develop stick-and-carrot measures to provide incentives for firms to behave opportunistically whilst adhering to a sustainable agenda.

To this end, climate change programmes have been or are being developed by all countries across Europe. The government remains the most influential party in supporting the achievement of sustainable targets through its position as the largest single client of the building industry and by using fiscal and regulatory measures to stimulate innovation and act as a broker in markets for environmental technologies. For example, in Sweden, the government subsidizes municipalities (many of which have their own energy firms) to implement measures that reduce the environmental impact, use energy more efficiently and promote the use of renewables and recycling (Kerr and Allen 2001). In Denmark, high electricity prices (maintained through the levy of additional energy taxes, including taxes to fund an Energy Savings Trust) have encouraged substitution of electricity for alternative energy sources to heat space and water, for example district heating (UNFCCC 1999). In Germany and Denmark, the government has legislated to guarantee price levels for electricity sourced from renewable energy (EEA 2001a). More generally, national policy has been geared toward improving the energy efficiency of buildings and (at a European level) the electrical efficiency of appliances (Kerr and Allen 2001). As Chapter 3 highlighted for Northern Europe, the imposition of more stringent building regulations has been the main tool used by governments to improve energy efficiency in buildings.

The government also has an important role to play as principal educator and disseminator of information to the industry and the general public and as market leader, prototyping innovative solutions through demonstration projects. Education campaigns to further awareness both in industry and among the public have been used effectively across Europe (for example, the Energy Efficiency Best Practice Programme for firms and the 'Are you doing your bit?' information campaign for households in the UK) in addition to improved fuel efficiency policies (for example, low sulphur fuel for use in high-efficiency boilers in Germany) and energy efficiency Eco-labels (for example, the increased market share of the most energy efficient products bearing the 'A' EU label in Denmark) (see Chapter 3; EEA 2001a, Kerr and Allen 2001).[5] The government can set a sustainable agenda and housing associations, responsible for commissioning social housing projects, can drive down the cost of adopting energy efficient technologies by, for instance, bulk purchasing technologies and using innovative procurement contracts to benefit from the creation of long-term relationships between designers, builders and suppliers. Also, repeating successful demonstration projects may help stimulate markets for sustainable building products, processes and services. An example of some of these measures can be seen in Scotland for example, where Scottish Homes, the national housing agency, assists approximately one-third of all house-building in Scotland. Scottish Homes has published a sustainable development policy (Scottish Homes 2000a) to inform housing associations' specifications, a sustainable housing design guide (Scottish Homes 2000b) to disseminate information about successful demonstration projects and a procurement advice note (Scottish Homes 2000c) to encourage closer collaboration with all actors in the building chain. In addition, Scottish Homes is currently consulting on the extent to which life-cycle costs should be considered alongside capital costs in the new Social Housing Grants (which replaced the traditional Housing Association Grants in 2002). In Chapter 5 we describe the most significant weakness of the strategy as the failure to evaluate adequately and repeat demonstration projects, further hindering the diffusion of innovative sustainable products and processes.

TECHNOLOGICAL CHANGE AND THE MITIGATION OF GREENHOUSE GASES

The Case of Active Solar Heating

It is widely accepted that technical change is an extremely important factor in addressing large-scale and long-term environmental problems such as climate change (Green et al. 2002, IPCC 1996, Weitzmann 1997). Technical change is seen as the cost-effective solution to mitigating greenhouse gases whilst maintaining economic growth. Given the long-term nature of climate change implications, high capital cost technologies combining medium term pay-back times with significant reductions in future greenhouse gas emissions have attracted financial support across Europe.

ASH offers a technological solution to the problems of carbon emissions and energy efficiency. Initial research and development and later market acceleration policies at the global level have been promoted through initiatives such as the Solar Heating and Cooling (SHC) Programme, established by the International Energy Agency (IEA) (an autonomous body within the Organisation for Economic Co-operation and Development (OECD)), within which countries collaborate to develop solar technologies to heat, cool, light and power buildings (Bosselaar 2001).[6] In Europe, a European Commission target of 100 million m^2 of solar collectors to be installed by the end of 2010 was announced in a 1997 white paper on renewable energies (RES 1997). The Soltherm Europe Initiative, an international collaborative project encompassing existing initiatives across 10 European countries, was subsequently established to install 15 million m^2 of solar collectors by 2004 by developing a framework of large demand satisfied by a sales and installation infrastructure (Van der Leun 2001).[7] Although active solar heating technologies can be seen as part of a bundle of technologies suitable for solar buildings (with passive solar design and active photovoltaic technologies), on their own they offer significant environmental savings (in terms of lower energy consumption) and, if one considers life-cycle costs/payback times, economic savings (in terms of lower energy bills).[8] With most (60 per cent) domestic water heating in Europe using natural gas potential savings in primary energy from 15 million m^2 of solar collectors amount to 1.19 million tonnes of oil equivalent (Van der Leun 2001). ETSU (1999a) estimate that where domestic water is heated by electricity, each kWh of electricity saved reduces carbon dioxide emissions by 0.585 kg. Thus, a 4m^2 solar collector on a single family house would be expected to provide between 1500 kWh and 2000 kWh per year, reducing carbon dioxide emissions by between 0.9 and 1.2 tonnes per year (ETSU 1999a).

An environmental focus from government, industry and the general public prompted considerable growth in the application of solar thermal technologies across Europe during the 1990s. In 2000, there were 100 manufacturers of solar collectors and other solar thermal products in Europe, employing over 13 000 people (see Table 4.1) with industry turnover of over £500m (Systèmes Solaires 2000). The structure of the European industry is characterized by many small and medium-sized enterprises and few large international firms, each of which accounts for similar levels of European market share.[9] Whilst the market for complete systems is domestically orientated, system components (such as absorbers and pumps) are traded internationally.

The European solar thermal market as a whole has grown by an average of 13 per cent since 1990 (Van der Leun 2001). Within Europe, there are wide variations in the size and growth rates of domestic markets, reflecting institutional factors such as government initiatives to stimulate the market (such as advertising campaigns and other dissemination strategies), private sector acceptance of solar energy and wider supportive public opinion – climatic factors are not important. For example, the widespread adoption of ASH systems in Germany has been facilitated by a combination of government (financial incentives) and industry (acceptance/promotion of solar from the traditional heating industry), assisted by an active national solar promotion programme (for example, 'Solar na klar') that has gained support from private individuals, local authorities and firms (Systèmes Solaires 2000, Van der Leun 2001).[10] By 1999, 8.8 million square metres of solar collectors were installed across the EU with Germany contributing over 30 per cent of the total (Systèmes Solaires 2000). In per capita terms, Denmark has the largest surface area of solar collectors per capita with 57m^2 per 1000 inhabitants, ahead of Germany (Eur'Observ'ER 2000) (see Table 4.1). Denmark, the Netherlands and Sweden also increased their total installed solar collector surface area year on year from the mid-1990s, whilst the cumulative surface area of collectors in the UK and France has fallen consistently over the same period.[11]

The technology used in solar water heating is well established (Bosselaar 2001). Although radical product improvements are still possible, the emphasis of innovation by construction-related firms is largely on the production process (for example, cost and price reduction through movement toward full-automation – production line assembly of ASH systems) and incremental adaptation of existing systems to meet more adequately customer requirements. There remains significant room for improvement in the integration of systems into different types of buildings and in the operation of different heating systems, with marketing and distribution economies to be obtained through better targeting. For example, individual houses constitute the largest sector of the market for solar water heaters, particularly in renovation rather than new build, and sales of systems in Germany and the

Table 4.1 *Solar thermal collectors across Europe*

	Total cumulative surface area by collector type (thousands m²) in 1999				Per capita surface area (approx. m² per 1000 inhabitants)	Industry Employment
	Glazed	Non-glazed	Vacuum	Total		
Germany	2130	400	220	2750	32	7500
France	321	332	5	658	11	120
Denmark	291	2	0.5	293.5	57	–
UK	132	75	–	207	4	–
Netherlands	116	90	4	210	14	200
Sweden	135	15	1	151	17	50
EU	7764	1549	–	9313	26	13070

Source: Systemes Solaire (1999), Systemes Solaire (2000), EurObserv'ER (2000), Bokhoven (2001).

Netherlands have been combined with the sale of high-efficiency boilers when old water heaters have been replaced.[12] For multi-house units or apartment blocks, there is an opportunity to integrate collective solar systems or district heating rather than local solar systems, which have been used widely in Scandinavian countries.

There are three types of technology common to ASH systems in Europe. Glazed solar collectors are the most common types of ASH system and the most common type of collector overall, accounting for 83 per cent of the total surface area of solar collectors in Europe in 1999 (see Table 4.1). Glazed flat plate water collectors work by pumping water between a transparent cover and a black plate with high thermal conductivity properties. Although there are many potential uses, the vast majority of glazed solar collectors are installed for individual water heaters: in 1999, 85 per cent of all installed glazed solar collectors installed were intended for individual water heaters. Other less common applications are for combined sanitary and hot water systems, accounting for 5 per cent of all glazed solar collectors, and thermal solar plants (where hot water is stored during the hot periods and used across local districts in cold periods), accounting for 1 per cent. The second type of technology, referred to as 'simplified collectors' or 'solar carpets', is used predominantly to heat water in outdoor swimming pools and accounts for 16 per cent of all solar collector applications. The third technology is vacuum solar collectors and is predominantly used in Germany for combined sanitary and hot water systems. Vacuum solar collectors, consisting of a set of evacuated glass tubes within which energy is absorbed and 'carried' to the water, are capable of carrying water of a higher temperature than other types of solar collector. Beyond Germany, vacuum solar collectors remain relatively expensive and Table 4.1 shows the weak diffusion of these solar technologies across Europe.

Comparison between countries is not straightforward since the systems differ considerably, for example, in terms of system design, average area of mounted collector and average size of water storage tank. These differences can be seen in Table 4.2.

Although costs and payback periods are difficult to compare, ETSU (1999b) estimated that ASH systems purchase and installation cost varied between 500 euros per m² and 1000 euros per m² across the three countries, with a payback period between 5.5 and 16 years depending on, for example, the level of solar fraction, the price of electricity and the usage profile. As Table 4.2 shows, the small Dutch ASH systems are the most efficient, in terms of energy generation and amount delivered to the storage tank per square metre of collector area, but their size means that they provide less hot water on a daily basis.[13] The large collectors in Sweden reflect the tendency for the system to provide space heating in addition to domestic hot water.

As referred to above, and as Table 4.2 shows, most ASH systems are used

Table 4.2 Comparison of characteristics and performance measures of ASH systems in Denmark, the Netherlands and Sweden

ASH system characteristics/properties	Denmark	The Netherlands	Sweden
Average system cost*	4580	1500	5045
Size of collectors and tank	5m² & 300-litre tank	3m² & 100–150-litre tank	10m² & 500-litre tank
Percentage of retrofit to new build	100% retrofit	70% retrofit	60% retrofit
Type of residences	90% single family	100% single family	88% single family
Average collector area (m²)	5	3	10
Ratio of collector area to storage area	19	26	11
Average annual solar irradiation (kWh/m²)	1031	991	1047
Solar fraction**	61	39	50
Average energy generated and stored***	392	643	331

Notes:
The number of systems in the ETSU survey varied between 14 in the Netherlands to 18 in Denmark.

* ASH purchase plus installation cost expressed in euros.
** Solar fraction refers to the level of household energy provided by the ASH.
*** Energy generated is measured in kilowatt hours per square metre area per year (kWh/m²/yr).

Source: ETSU (1999b).

in individual houses and tend to be introduced during renovation when heating systems are replaced. Incorporation of ASH systems into the traditional heating industry has benefited Germany and is providing the basis for further market growth in the Netherlands in combination with national schemes such as 'BelDeZon' (Call the Sun) and 'Ruimte voor Zonnewarmte' (Space for Solar) (Van der Leun 2001). Since the market for ASH is geared toward individual households, the payback time is very important, though decisions to adopt an ASH system will also depend on factors such as environmental awareness, comfort and status (Bosselaar 2001). With an emphasis on providing a cheaper, more widely available technology the government has an important procurement role to play, creating markets and encouraging innovation by acting as a broker in a 'technology procurement' strategy. Across Europe there has been no shortage of innovative projects using solar thermal technologies in the public housing sectors where it has been common for social housing to prototype ASH technologies and act as demonstration schemes, largely funded by the EU, national governments, local authorities and housing agencies. For example, under the European Commission's THERMIE programmes, Solar Housing through Innovation for the Natural Environment (SHINE) and Solar Urban New Housing (SUNH) projects, ASH systems have been included in social housing projects across Europe:

- in Reading, Greenwich and Portsmouth in the UK by Reading Borough Council, Hyde Housing Association and Sovereign Housing Association;
- in Amsterdam, the Netherlands by Partimonium Housing Association;
- in Gardsten, Sweden by Gardstens Bostader;
- in Engelsby, Germany by BIG Heimbau; and
- in Farum, Denmark by Boligselskabet Farumsodal.

Both the SUNH and SHINE programmes are concerned with demonstrating new energy efficient technologies and achieving significant energy savings (above 50 per cent) in social housing across Europe. In Scotland, where one could be forgiven for thinking that the climate would inhibit the use of solar technologies, there have been many successful demonstration schemes for new-build and retrofit housing funded by the EU and Scottish Homes, the National Housing Agency. In fact, solar radiation levels are not significantly different from those experienced by southern England (between 4 and 4.4kWh/m^2 per day in July in Scotland compared to between 4.8 and 5kWh/m^2 in southern England), and because of better air quality, solar transmittance tends to be better (Everett 1996, MacGregor 2000).[14] In addition, relative to the UK, Scotland has a large number of houses without access to gas or with only electric heating and many which are difficult to heat and/or insulate and which are prone to dampness and condensation (MacGregor 2000). The EU has funded three Ecorenewal projects in Glasgow using active solar heating as part of an ecological retrofit of a Glasgow City Council three-storey tenement building, a Meadowside and Thornwood Housing Association four-storey tenement building and Williamsborough Housing Association four-storey tenement. In addition, other sustainable demonstration projects, such as that by Shettleston Housing Association, with funding from Scottish Homes, have used ASH technology to provide domestic hot water for 16 houses through a combination of geothermal technology (from a nearby disused mine) and ASH collectors.[15]

CONCLUSION AND POLICY IMPLICATIONS

High cost and unknown cost effectiveness have contributed to the slow adoption of ASH, in addition to a general lack of knowledge about the products and their suppliers and concerns over their reliability and safety. The high costs of ASH systems in the UK can be attributed to the aforementioned market distortions, which do not take account of the environmental costs of conventional energy sources, exacerbated by the relative small size of the industry, increasing the per unit marketing/selling costs and limiting production, managerial and inventory economies.[16] The need to include new technologies in housing is not important simply because of the need to construct high-quality, low-cost housing in the shortest possible time. Environmentally friendly or energy-efficient technologies are particularly attractive for social housing developments since they reduce energy consumption (and hence energy bills) helping to combat fuel poverty and are affordable when life-cycle costs are translated into slightly higher rents paid over a number of years by one or more tenants.[17] The housing sector, both private and public, sidesteps another important barrier hindering innovation in the building industry – the one-off nature of building – through the capacity for multi-unit schemes. Multi-unit schemes also unlock economies for the bulk purchase of sustainable technologies (such as ASH systems) through lower marketing and manufacturing costs per unit, more effective installation management and faster installation times. The cost of ASH systems is cheaper when purchasing in bulk and housing associations are one client in a position to benefit from significantly lower unit costs. ETSU (1999a), in a study of public sector housing associations and private sector housing developers, identified the principal barriers to the adoption of ASH systems as the cost, the lack of available information on the long-term costs, the perceived energy rating benefits and concerns over the aesthetic qualities.[18] Yet the social housing sector has an added (and arguably a more important) incentive to install solar technologies because of their capacity to alleviate fuel poverty by reducing energy bills. National housing agencies, such as Scottish Homes in Scotland, could help housing associations through increased HAG grants to those projects where particularly innovative technologies or construction methods are used to increase energy efficiency or reduce the environmental impact. More consideration needs to be given to life-cycle costs, perhaps in line with an anti-poverty strategy or where rents could be raised to cover higher capital costs (for example, where tenants benefit from lower fuel bills). Alongside national campaigns, local awareness initiatives could be funded. For example, because the solar technologies are designed by consultants to be relatively easy to install, solar clubs could be established where volunteers explain to sub-contractors, such as plumbers, how the system works (Gilbert 2000a). Also, councils can be encouraged to promote solar technologies though audits.[19]

The government can address the problems of market distortion through regulation, taxation and legislation and can encourage innovation by acting as a broker in a 'technology procurement' strategy. In addition, government may introduce grants and fiscal incentives, channelling funds towards R&D and facilitating economies of learning and experience, beginning with demonstration projects and continued through information dissemination. Regulation and legislation at both the local authority and national level can promote solar technology. At the national level, the UK could follow the example set by Denmark, Germany, the Netherlands and Sweden. Fixed-price schemes for renewable technologies are marketed in Denmark and Germany and direct capital grant support and tax incentives for renewable energy projects are provided in Denmark, Germany, the Netherlands and Sweden. Denmark and Sweden also offer net metering to encourage small-scale renewable energy production (Thorp 2000). Some European countries offer low-interest loans for solar water heating and others, such as Norway, offer lower rate mortgages to buildings that will improve the quality of the built environment (for example, energy-efficient buildings, healthy housing) (Gilbert 2000b, Thorp 2000). Governments in the UK and France could replicate these policies. Support for private homeowners or tenants could be in the form of access to low-interest loans for using solar technologies in their domestic dwellings. Support for industry could be provided in the form of tax incentives available to firms using renewable resources and retrofit improvements could be entitled to retrospective additional capital cost rebates following an audit.

Because unlike many other industries, innovations in construction are not implemented within construction firms themselves but on the projects on which firms are involved, (Gann 2000, Winch 1998) the management of innovation in construction is complicated by inter-firm coordination and demands negotiations along the building chain. Therefore manufacturers and suppliers of ASH systems can not be considered independently from the building chain. Previous research has shown that explicit consideration of implementation activities in construction firms can significantly improve both the innovations and the degree to which they can be used effectively within the construction industry (Slaughter 1993, 2000).

Housing associations can be important catalysts for the use of ASH systems through clear specifications for their implementation by designers and contractors, rather than simply accepting the standard technologies. At present the design of buildings to include ASH systems is a specialized skill, offered by building service engineers and architects with particular expertise. So too is the skill required to mount the ASH systems on the roof/façade and maintain the systems (though the electrical and plumbing side is intentionally kept relatively simple). Just as architects and consulting engineers can develop a market niche in the design of ASH buildings, contractors that adopt ASH systems can also develop a reputation for their willingness to try new products and can increase their markets for projects including sustainable technologies (see Chapter 5). By raising the visibility of the adoption of this sustainable technology, construction firms can benefit their organizations as a whole from the reputation in obtaining and installing this innovation, while confining the risk to this specific technology (Slaughter 2000). When the market is more developed, a commercial advantage may be conferred on those construction firms in the vanguard of ASH systems. Governments wishing to encourage the further use of ASH systems need to acknowledge the difficulties in the implementation process and the need for learning about this technology across the construction industry. An important area for government action, therefore, is in increasing the capacity of construction firms to identify appropriate sustainable technologies and evaluate their potential costs and benefits.

NOTES

1. Although in the EU, domestic sector energy consumption per square metre fell by 8 per cent between 1985 and 1997, final energy consumption increased steadily, rising 4 per cent between 1985 and 1998 (EEA 2001a, EEA 2001b). The number of households increased by 19 per cent between 1980 and 1995; the average size of dwellings (in m²) increased by 5 per cent between 1985 and 1997; the average number of persons per household fell by 12 per cent between 1980 and 1995; and electricity prices fell by 1 per cent per year between 1985 and 1996 (EEA 2001a, 2001b, ENERDATA/Odyssee 1999).
2. During the 1990s, policy initiatives focused on improving inter-organizational co-operation and sustainability by changes in public funding and in the organization of production based on the recommendations of two important government reports (The Latham ('Constructing the Team') Report in 1994 and The Egan ('Rethinking Construction') Report in 1998). Sustainable initiatives, aimed at reducing energy consumption, have also been promoted at the national level (for example, UK Climate Change Programme, Building a Better Quality of Life: A Strategy for more Sustainable Construction, (Energy Efficiency Standards of

Performance (EESOP)) and local level (for example, Home Energy Efficiency Scheme (HEES), Home Energy Conservation Act (HECA)). For example, appointing an integrated design team at an early stage in the construction of 'green buildings' (Sorrell 2001) and using alternative procurement strategies to address sustainable development issues (such as higher environmental standards, eco-design principles and life-cycle implications) (Pollington 1999).

3. In Chapter 5 we discuss the advantages and disadvantages of alternative contractual arrangements from the point of view of housing associations, architects, consulting engineers and contractors operating in the Scottish social housing sector. Unsurprisingly, the architects and engineers on the one hand and contractors on the other have opposing views on the relative advantages of traditional and design-and-build type procurement forms.

4. In Chapter 5 we show how a desire to be more sustainable, in this case a Scottish housing association's desire to use boron-treated timber as opposed to traditional chemical-treated timber, is compromised by the high-pollution-cost transport required to source the timber.

5. 1992 EU council directive 92/75/EEC introduced eco-labelling, grading electrical appliances between A (the best, most efficient) to G. By 1999, the sale of most Grade E, F and G refrigerators and freezers had been banned (EEA 2001b).

6. Fourteen European countries and the European Commission are involved in the SHC programme agreement alongside Australia, Canada, Japan, Mexico, New Zealand, and the USA. The research is task-based with individual countries funding and conducting their own work within particular tasks.

7. Soltherm Europe Initiative includes partners in Austria, Belgium, Denmark, France, Germany, Greece, Italy, the Netherlands, the UK, and Spain in addition to a number of pan-European partners.

8. Van Zee (1999) describes the integration possibilities for solar thermal collectors at the design stage giving examples from Dutch solar collectors with a dual function, acting as a shading device, as the roof of a sun lounge, as façade cladding and as a sun porch amongst others.

9. The European Solar Industry Foundation (1995) estimated that only 15 per cent of firms employ over 30 people over Europe. In Germany, for example, small firms with turnover below €2.5 million, and large firms, with turnover in excess of €5 million, share 41 per cent and 43 per cent of the total European market respectively. Medium-sized firms, with revenue between €2.5–5 million account for only 16 per cent of market share (Systèmes Solaires 2000).

10. The 'Solar na klar' programme was initiated in 1997 by Baum, a group of green entrepreneurs representing small- and medium-sized solar firms, and has federal state financial backing from Gerhard Schroder, the German Chancellor and Jurgen Tritten, the Environment Minister. The campaign has raised public awareness through advertisements and PR work – 65000 people requested information in 2000 – and is funded by the private sector (for example, the solar industry and other private firms) and public sector (for example, federal and state funding) (van der Leun 2001).

11. The reduction in the cumulative total can be partly explained by a higher number of systems removed due to obsolescence than the number replaced. The UK and France were particularly active during the 1980s in installing solar collectors but these are now approaching the end of their lifetime.

12. Over one half of solar systems sold in German in 2000 were sold in combination with a new boiler (Van der Leun 2001).

13. The Dutch system differs from all the others since it uses a drainback system protect against freezing; the other country's systems used glycol (ETSU 1999b). The small average storage tank sizes in the Netherlands can be seen by the high ratio of collector to storage size; adding to the efficiency of the system by maximizing heat transfer between the fluid loop and the refilled cold water storage tank every time the hot water is used.

14. Note: $5 kWh/m^2$ is enough energy to heat an average bath of hot water.

15. For more information regarding these examples, see www.ecorenewal.com, Scottish Homes (2000b) and Gilbert (2000a, 2000b). See also Chapter 5 for further analysis of the innovation

process in the Meadowside and Thornwood Housing Association, Williamsborough Housing Association and Shettleston Housing association projects.

16. The UK is one of the domestic markets in Europe where production exceeds demand and exports constitute a considerable percentage of industry income. In 1994, UK production equalled 7 per cent of total EU production despite the UK market only accounting for 2 per cent of the total number of systems in the EU (see Table 4.1).

17. It is estimated that 37 per cent of households in Scotland may be experiencing fuel poverty (Scottish Homes 2000a).

18. Of fifteen associations approached (2.5 per cent of all UK housing associations), seven replied, only two of which were enthusiastic about the potential for ASH systems. Of nine housing developers approached (12.5 per cent of all UK housing developers), two replied, only one of whom requested more information and expressed an interest in bulk-buying ASH systems.

19. For example, in the UK, councils can promote solar use through policies such as Agenda 21 and HECA in addition to their housing strategy (for example, energy efficiency of the housing stock) and anti-poverty strategy. Councils have additional incentives to promote solar technologies in response to external audits to determine their 'environmental stewardship' (for example, a council's energy and water strategies and commitment to renewable energy) and through the Environmental Management and Audit Scheme (which assesses the environmental impact of the council's actions and the actions of its customers and clients) (Huskinson 1998).

5. Networks and sustainable technologies: the case of Scottish social housing

INTRODUCTION

A number of recent studies reveal interest in networks as particular organizational forms that facilitate innovation. Different contributions have analysed the 'make or buy' decision and it has become increasingly clear that alternatives to internalizing are often found in some 'third way' (rather than market or hierarchy forms of organization) including joint ventures, networks or clans (Buckley and Casson 1990, Miles and Snow 1986, Ouchi 1980, Pfeffer and Nowak 1976). The general argument is that networks may create a high sense of mutual interest, communication and participation among organizations that may facilitate the efficient processing of information and generation of knowledge (Castells 1996, Nohria and Eccles 1992).

As argued in Chapters 1 and 2, the construction industry is particularly well suited for the examination of these inter-organizational relations because it can be regarded as an archetypal network system where a coalition of organizations – including contractors, the government, clients, designers, sub-contractors, suppliers and tenants – come together on a temporary basis to undertake each project (Gann 2000, Winch 1998). This chapter assesses the case of the introduction and diffusion of sustainable technologies in the Scottish social housing sector. This case is of particular interest because since 1997 Scottish Homes, the National Housing Agency, and the Scottish Federation of Housing Associations, have been active in fostering inter-organizational collaboration and promoting sustainable construction products and processes through issuing policy guidance, briefing notes and training schemes.

However, many of the problems of the performance of the construction industry seem to stem from inadequate inter-organizational co-operation. This chapter aims to shed light on the interactions and interdependencies between organizations, embedded in a tight network of production relations, which have an important role in shaping the process of sustainable production and innovation in the construction sector. In particular, the chapter shows the contradictions between policy aims to promote the implementation of sustainable technologies and the organizational relationships in the social housing sector that appear to militate against the achievement of these objectives. The chapter assesses the role of the national housing agency and the local housing associations responsible for acting as promoters of new build and retrofit social housing projects and their interaction with other organizations in the building chain. Our analysis is informed by extensive semi-structured interviews with representatives of ten housing associations, the national housing agency and the housing associations' trade association, three private sector housing developers, five architects, three consulting engineers and three contractors operating in the social housing sector mainly in and around Glasgow and Edinburgh (see Table A.4 in the Appendix for a list of organizations interviewed).[1] All organizations interviewed had some experience of working on social housing projects that included some sustainable elements. The projects included sustainable technologies or were built according to a sustainable design (for example, projects used natural cavity wall insulation or had passive solar design). Where organizations had been involved in demonstration projects (within which more money is available to prototype new sustainable technologies), more high-tech options, such as active solar heating and geo-thermal heating, were used.[2]

This chapter is organized as follows. The first section assesses the extent to which innovation studies have considered inter-organizational relationships and explores the particular case of construction industry innovation. The next section gives an overview of issues surrounding the implementation and diffusion of sustainable technologies in the Scottish social housing sector. The third section considers the role of the relations of housing associations with contractors, architects and consulting engineers in facilitating the implementation of new sustainable technologies. The fourth section assesses the implications of the study for the promotion of collaboration and sustainability. Conclusions follow.

INNOVATION STUDIES AND ORGANIZATIONAL RELATIONSHIPS

Innovations are developed through many co-ordinated and contributing organizations. However, much of the innovation literature, particularly that within management studies, has placed the individual innovating firm at the heart of the analysis. This is not to suggest that this literature has ignored the fact that many firms involve other organizations in production and innovation.

The early work by von Hippel (1988) and Lundvall (1988) has stressed the importance of relations with users and suppliers in the innovation process. Nevertheless, the innovation literature has over-emphasized the significance of strategic choices of individual firms or of firms and their bilateral relations with other firms, suppliers and users.

Little attention has been placed on the interactions and interdependencies between organizations, based on enduring and socially embedded relations, which have an important role in the process of innovation. For example, the 'national systems of innovation' literature (Freeman 1987, Lundvall 1992, Nelson 1993) deals with the network of institutions in the public and private sectors the activities of which contribute to the introduction, import and diffusion of innovation. However, it pays little attention to how these different institutions and organizations interact. The 'sectoral systems of innovation' literature (Edquist 1997, Malerba and Orsenigo 1996b) has made some progress in analysing the relation between organizations in a sector and its supporting institutions. However, these analyses tend to concentrate on new scientific and technological knowledge, and have little interest in more mature manufacturing sectors and services.

These problems are all the more relevant for the construction sector, in which the production and innovation process of firms is embedded in a tight network of organizations which includes other industries in the supply stream, end-users as well as the government. Some recent contributions in the innovation literature have explored the suitability of 'project-based' organizational forms in the production of complex products and systems (Hobday 1998). As in other project-based sectors, the project-based nature of work in construction implies that firms have to manage networks with complex interfaces, involving many organizations from a range of sectors, temporarily working together on project-specific tasks (Gann and Salter 2000, Winch, 1998). However, performance and competitiveness in the construction industry does not depend solely on the single firm, but on the efficient functioning of the whole network.

Moreover, the pressure to meet the new demands for sustainable technologies or processes presents another important challenge for the network of organizations involved in the construction process. As noted in Chapter 4, the nature of the construction industry, the characteristics of the final product and the operating environment hinder significantly the adoption of innovative products and processes. Traditional contractual arrangements that bind collaborative parties within construction projects have exacerbated the problems inhibiting the identification and implementation of new products and processes through mutual distrust, lack of communication and time and cost constraints. These barriers to the adoption of innovation are particularly acute when one considers the use of sustainable technologies or processes, typically characterized as high cost, high risk and about which less information is available (see Chapter 4).

Although the above may explain the barriers to the adoption of more sustainable construction processes and energy-efficient technologies in the domestic private sector, in the case of social housing, those parties involved in commissioning and building have more holistic aims. Housing associations tend to have a more practical approach to sustainability, seeking to address fuel poverty and achieve comfortable living conditions for tenants. However, even if the concern is of a practical nature, the main barrier to using sustainable construction methods and sustainable construction materials remains the often higher capital costs.

Over the last decade, the concepts of improved inter-organizational co-operation and sustainability have been at the forefront of policy initiatives in construction. The need to appoint an integrated design team at an early stage in the construction of 'green buildings' (Sorrell 2001) and the need to use alternative procurement strategies to address sustainable development issues (such as higher environmental standards, eco-design principles and life-cycle implications) (Pollington 1999) are well-established concepts. In the construction industry, issues of greater collaboration and procurement have been addressed through changes in public funding and in the organization of production based on the recommendations of two important government reports (The Latham ('Constructing the Team') Report in 1994 and The Egan ('Rethinking Construction') Report in 1998). Sustainable initiatives, aimed at reducing energy consumption, have also been promoted by government policies, regulation and economic instruments at the national level (for example, UK Climate Change Programme, Building a Better Quality of Life: A Strategy for more Sustainable Construction, EESOP) and local level (for example, HEES, HECA).[3]

Scottish Homes, the National Housing Agency in Scotland, has been at the forefront of policy development and therefore represents an interesting case study. Scottish Homes has encouraged collaboration and sustainable construction by housing associations through policy documents and advice, through funding demonstration projects (and repeat projects) and through monitoring, collating information and disseminating results (for example, Scottish Homes 2000a, 2000b, 2000c). The Scottish Federation of Housing Associations also has an important education role and has been active in providing best practice guidance and up-to-date information and training (for example, SFHA, 1999).[4] The next section gives an overview of issues surrounding the implementation and diffusion of sustainable technologies in the Scottish social housing sector.

THE SCOTTISH SOCIAL HOUSING SECTOR AND SUSTAINABLE TECHNOLOGIES

Climate change provides a strong motivation behind the use of more sustainable building products and processes. The domestic sector has the capacity to reduce greenhouse gases emissions significantly by increasing energy efficiency and reducing energy demand from non-renewable resources. Sustainability has been the focus of recent Scottish Homes policy documents (for example, Scottish Homes 2000b) proposing increased thermal performance of housing (through increased minimum SAP (Standard Assessment Procedure) ratings for new-build and retrofit housing), reduced consumption of physical resources (through minimization of energy and waste and maximization of recycling, local sourcing and the use of low-embodied energy and renewable materials), encouraging brownfield sites (through additional funding) and educating the consumer about more sustainable lifestyle choice. Housing associations recognize the importance of these issues (notwithstanding the fact that their future funding is dependent on achieving certain sustainable criteria) but aim mainly to provide high quality affordable social housing and to eliminate fuel poverty rather than, in the words of a housing association representative, taking a 'high-brow, ecological design for ecological design's sake' approach.

Despite policy aims to promote the use of more sustainable building products and process, one of the main barriers to the adoption of sustainable construction methods and energy efficient materials is their cost. This is not simply a reflection of the problems of fragmentation in the industry, the risks associated with introducing new technologies and market imperfections (which do not take account of the environmental and social costs) (see Chapter 3; Gann 1994, Nam and Tatum 1988). In addition, there are specific problems with sourcing innovative materials in niche markets with few specialist suppliers, particularly when suppliers are overseas. Also, when suppliers fail to deliver, significant design alterations are required, often compromising the initial energy efficiency. One of the architects we interviewed provided the following illustration:

> The issues on the design side that caused us frustration were generally having identified manufacturers and suppliers of [suitable] products and having been promised by their representatives that they could deliver, when we got on site, all of a sudden we couldn't get them in this country. [For instance], we [found] a glazing system [in] Germany which we were told we could get through Scottish agents but a few weeks before we went to tender they suddenly said, 'No, it's a policy decision we're not going to import them to the UK'. ...So it was a combination of finding alternative products, not being able to find the exact matches and having to redesign and to incorporate what was available. And the [ultimate] glazing system is significantly compromised from the original design.

Supply-side problems also raise questions about the 'true' sustainability of the products. The desire to use low embodied energy materials must be weighted against the performance of the material (as shown in Chapter 3) and against the practical environmental costs of sourcing and transporting. An example was provided by a representative of a housing association:

> There were only two places where we could get [the timber] treated – Aberdeen and Inverness – and Inverness closed down just as we were about to send the timber off. In actual fact, I think you'll probably find that the transport costs to get the timber treated are more expensive than the timber since it had to go to Aberdeen and because they would only take one lorry load at a time. We tried to use boron treatment, which is healthy [when compared] to traditional chemical treatment, but if you look at the whole cycle of ecological issues then, arguably, we're imposing more environmental damage with greater transport costs.

Also, because there are steep learning curves associated with the use of sustainable technologies and because housing associations are engaged in few innovative projects, there is a lack of in-house expertise in using these sustainable building products and processes and, therefore, additional costs must be incurred buying in the expertise from outside. Despite the requirements of the Home Energy Conservation Act, because energy management requires wide and disparate knowledge, local authorities fail to provide the housing associations with adequate advice and support. Thus, all these factors add up to an inherent conflict created by the (usually) more expensive use of sustainable technology and the external funding limits set by Scottish Homes and ultimately the Scottish Executive. For the housing associations, the costs of negotiating the introduction of sustainable technologies in housing often meant that short-term solutions were more attractive in terms of meeting tenants' needs:

> If you're working with people who have awful accommodation and you know that you can deliver a project that will be warmer, better for them, then there is a huge incentive to push in that direction. Why go and put solar panels on, for instance, and have all the arguments and all the hassle that this causes in terms of politics. You've got people sitting in accommodation they can't heat and I'm not willing to sacrifice people's lives more than I have to just to argue the point that if Scottish Homes wanted to they could agree to fund renewables.

Some initiatives have attempted to address this conflict between the higher costs of sustainable technologies and the funding limits. Indeed, the principal aim of housing associations in Scotland is to provide low-income tenants with low-maintenance easy-to-manage housing that produce relatively low energy bills. The capacity to pass on savings to tenants in the form of lower energy bills provides a stronger incentive to use energy efficient technologies. Nevertheless, the application of sustainable technologies is hindered by the emphasis placed on capital costs. In the words of one of the architects interviewed:

> If something can't be paid off in about five years then it's not cost effective. There is no such thing as life-cycle costs.

However, there have been some improvements. Over the last few years there has been a growing appreciation of 'life-cycle' costs, where capital costs are considered alongside costs incurred over the building's lifetime (for example, maintenance costs, availability of materials, installation costs and forecast lifespan) or alongside environmental and social costs (for example, externalities generated in the production and use of the materials, liability and risk issues). Considerations such as these lend themselves to both a better environmental and a better social agenda (predominantly if the payback time is not too long and the energy bills of tenants are reduced) but tend to get crowded out by capital costs. Although 'payback' calculations are not explicitly recognized in mainstream funding schemes, representatives of three of the housing associations interviewed argued that they have recouped higher capital costs through higher rental streams. Also, tenants of these initiatives of housing associations have benefited from significantly lower fuel bills. For example, in the Shettleston Housing Association's Glenalmond Street development, the rents increased marginally but the tenants were provided with virtually free heating and hot water all year round by a geothermal district heating system supported by solar power and extremely high levels of insulation. Representatives of other housing associations interviewed argued that it remained difficult to consider payback costs either because they wanted to keep rents at affordable levels or because they had no mechanism by which to charge higher rents to compensate for lower fuel bills. When considering payback costs, another important consideration was the distinction between 'high-tech' and 'low-tech' solutions. As argued by the representative of a housing association:

> Some things in a sustainable development can be done without too much expense. For energy efficiency and sustainable development there are high-tech and low-tech options. The former requires a lot of senior officer input at the design stage and continuous senior level input on the management and maintenance side. Housing associations are well-positioned to install it, monitor it and provide feedback, but there's no way, with the [limited] private finance that we can raise and the Housing Association Grant that we receive from the Government, that we can afford a high-tech option.

For example, on a project including passive solar design and the use of sustainable resources (for example, borate treated timber, natural water-based paints) and other energy efficiency measures (for example, higher insulation, condensing boiler, low energy lighting), capital costs increased by 25 per cent per unit, up to half of which was accounted for by the passive solar system. All housing associations interviewed expressed an interest in implementing new sustainable technologies but felt more strongly about improving the overall living conditions of their tenants, preferring to channel funds toward 'low-tech' energy efficient options such as natural insulation (for example, Warmcell) and double glazing. A representative of a housing association said:

> The climate that we have here is cold, damp [and] miserable at times so of all the things we can do ... heating and insulation are the two fundamental things. We're looking at things like warm cell insulation and standard condensing boilers, low-flush WCs – all standard technologies. We avoid the high profile demonstration projects that you're going to do once. What we'd rather do is achieve a standard [specification] that we can put into all the properties and then just look to continually improve that standard. We used solar panels many years ago in a tenant scheme and it went badly wrong. Instead we prefer to be in a position to do it in all properties rather than just to do it in one or two to see how it works.

In general, there is a consensus among all parties that higher capital costs should be weighted against payback times and the benefits to tenants and the wider environment. For example, an architect interviewed argued:

> [Funding should be available for] higher capital cost projects whenever it can be demonstrated that the cost is reflected by a payback time [that] justifies it. I'm not making any judgement as to whether the payback time should be two years or 20 years – that's a specialist field – but I do feel that on any mainstream project there should be an allowance for an increase in capital costs for construction, so long as there is a benefit over a certain period of time.

Although the housing associations are principally interested in alleviating fuel poverty, one more unexpected consequence of introducing energy efficient measures has been the positive spillovers in terms of improved quality of life experienced by the tenants. For example, a consulting engineer argued:

> Where solar ventilation has been used in the houses, whilst you can't justify doing the solar ventilation on the basis of energy cost, they feel much drier and warmer and more comfortable than a typical tenement course which is usually a fairly dank, cold, unwelcoming place. So there is a quality of life element, which I think is being appreciated by the tenants.

Indeed, tenants also have an important role in supporting the diffusion of technologies. Users' experience provides an important gauge for the success of new technologies prototyped in housing schemes and should not be underestimated since the success of innovation, in terms of its diffusion, depends on whether, following implementation, the operational complexity does not hinder or prevent its correct application. The potential problems are particularly acute when considering sustainable technologies since they are associated with energy efficiency and present quite a radical change in the way people heat their homes, replacing traditional incumbent systems such as centrally situated gas fire.

Therefore, Scottish Homes has been active in policy initiatives to encourage sustainable technologies. However, problems with the sourcing of innovative materials in niche markets and consideration of capital costs conflict with the achievement of policy aims.

INTER-ORGANIZATIONAL RELATIONSHIPS AND SUSTAINABLE TECHNOLOGIES

The project-based nature of construction suggests that the implementation and diffusion of innovation in the construction industry require the participation and collaboration of the many different parties involved in the construction process. Each of the parties may have their own distinct roles and responsibilities for encouraging innovation but it is the relationships and interactions with each other that determine the success of innovative projects (see Chapters 1 and 2).

Scottish Homes provides much of the funding for conventional housing schemes and innovative competition schemes in Scotland.[5] Competition projects offer additional funding in order to design social housing to a higher specification. In 1997, Scottish Homes promoted a competition requiring teams of housing associations, architects, engineers and contractors to specify an alternative, innovative sustainable solution to social housing – using sustainable design (for example, passive solar), sustainable products (for example, natural thermal insulation) and sustainable technologies (for example, active solar heating).[6] As argued by Scottish Homes, the real success of competition schemes may be measured by the diffusion of innovative ideas or technologies across other conventional projects. Innovative schemes are most useful if they are replicated because subsequent projects can benefit through economies of learning and experience (for example, in both the design and construction process, in the sourcing of new materials, employment of sub-contractors, etc.) and through improved trust between the key participants. The failure to repeat (some elements of) demonstration projects may hinder the further diffusion and application of new technologies.

Each housing association provides a specification document outlining the specific design and function criteria that buildings must meet. The Scottish Federation of Housing Associations provides training and advice on project briefings (see SFHA 1999) and Scottish Homes provides broad specification outlines. But the housing associations then have the autonomy to tailor the specification to their satisfaction and the specification document is under constant review and evolution. Of course, one could argue that this precise and explicit specification document may hinder radical innovation whilst promoting incremental change. Nevertheless, alterations to the specification document or design guide is an important ingredient in replicating technologies prototyped in demonstration projects.

This section explores the contradictions between policy aims to promote the introduction of sustainable building technologies and processes and the barriers imposed by inter-organizational relationships in the construction industry. Although a number of initiatives in the construction industry have promoted closer inter-organizational relations with the aim of facilitating the successful implementation of innovation and especially sustainable technologies, there remain important tensions and contradictions between the interests of the different parties involved in the construction process that may militate against the achievement of these objectives.

Relations between Housing Associations and Contractors

Traditionally in the UK, contractors are selected by competitive tendering on a lowest cost basis and the relationship with the client (the housing association) is characterized by a lack of communication, trust and co-operation (Miozzo and Ivory 2000). Surprisingly, a number of the housing associations admitted engaging in 'informal' long-term relationships with contractors, architects and engineers for many years. More recently, as alternative procurement forms have become more popular, the relationship between housing associations and contractors has become more important. Given that housing associations are investing public money they tend to feel a responsibility to employ local contractors, architects, engineers and sub-contractors whenever possible, fuelling a regional multiplier effect, contributing to local employment and providing local firms with the stability to organize apprenticeships. Local contractors also tend to have a core number of tradesmen on their books, the quality and reputation of whom are known to the housing association. Representatives of housing associations interviewed argued that the size of the contractor was important, most having informal long-term relationships with four or five local medium-sized contractors. For example, one argued that:

> We target the size of contractors we work with. I don't particularly want to work with [a large contractor] because I don't believe the scale necessarily provides the discount. Large contractors coming into a small market are easily bored, hard to work with and difficult to get decisions from. Medium-sized contractors are more eager and anxious to learn and to please. You need to be able to speak to the person in charge, who's responsible for making the decisions. I am able to speak with the

managing directors of all my contractors and I know they will keep in contact because I am a key part of their finances – we pay within 21 days, so we're good for cash flow.

Nevertheless, there are tensions between housing associations and contractors, reflecting their contrasting non-profit and profit motives (and, more generally, low industry profit margins), which impact adversely on innovation. In the words of a representative of a housing association:

> With fixed costs the housing association wants the best product possible from the contractor. The contractor, however, although prepared to negotiate on areas of specification (for instance, in a design development project) is only interested if it increases their profits. They're not going to suggest a project that would be less profitable. And the problem with an innovative project is that they are generally less profitable.

Associated with the support of long-term relationships between housing associations and the building chain are changes in the method of procurement. Over recent years, reflecting a relaxation of Scottish Homes funding and procurement restrictions, housing associations have had more choice of contractual arrangements. Although all the housing associations interviewed argued that the vast majority of contracts are still procured through a traditional tendering route, many housing associations have prototyped innovative procurement forms, including traditional 'off-the-shelf turnkey', 'design and build' and 'negotiated design and build' (mentored partnering) more recently. The adoption of alternative procurement methods has been gradual, with housing associations initially engaging with 'trusted' contractors. Even with good client–contractor relations, however, experiences of the housing associations engaged in these alternative contractual arrangements have been mixed. Benefits in terms of cost certainty (guaranteed price) are tempered by the loss of control. In addition, housing associations offered contrasting evidence on whether 'design and build' delivers projects faster.

In terms of innovation, the representatives of housing associations interviewed believed that long-term relationships (combined with alternative procurement forms) helped overcome conservative tendencies, increased trust between the parties and encouraged the adoption of new technologies. In the words of an interviewee:

> Some contractors are very conservative whilst others are much more innovative and will volunteer to use new technologies; 33 per cent of our contracts are traditional and 66 per cent are design and build; probably for the reasons of being able to use new technologies in continuing partnerships.

'Design and build' contracts allow the contractor to be involved earlier in the building process. All representatives of housing associations interviewed acknowledge that there are benefits from the contractor's earlier presence, for example, to improve the buildability of the scheme and to add the contractor's construction expertise to the design and specification detail, the programming and the site management and control. Also, without the contractor's presence the design team may overlook or give less priority to issues that may have a significant impact on cost, such as the contractor's space requirements. For example, the representative of the housing association in one of the developments thought that many of the problems (for example, sourcing of materials, selection of sub-contractors) could have been avoided if the contractor had been involved earlier, before the detailed system design. Even housing associations with more reservations about the motivations of contractors acknowledge the advantages of their early involvement, as shown in an interview with a representative of a housing association:

> The only benefit of bringing a contractor in early is if there are parts of the design specification where the contractor can offer you cost efficiencies. Then you can use those savings to be more innovative elsewhere.

The early involvement of contractors is regarded as advantageous by housing associations and contractors alike. However, the non-profit and profit motives of housing associations and contractors respectively remains an important source of tensions and conflicts of interest.

Relations between Housing Associations and the Design Team

Another problem in attempting to build closer relationships with the aim to promote sustainable technologies is that not all parties in the construction process have welcomed the adoption of alternative procurement strategies, particularly in terms of innovation and sustainability. As one would perhaps expect, the architects and consulting engineers interviewed almost exclusively considered that innovation was facilitated by the traditional contract form and stifled by 'design and build'. For example, a consulting engineer interviewed argued:

> Under a 'design and build' arrangement, the contractor, having usually taken the contract at a price and adhering to a tightly controlled brief, does not have much room. Nor does the engineer if he comes in to detail the design work. In that extent, it stifles innovation.

An architect interviewed also argued the following:

> Adoption of technologies or innovation, I think can only be generated by the more traditional forms of contract. If you have a full remit then you can explore innovation: putting to the client, considering the cost implications [before] going to

tender. On a 'design and build' basis it's far less likely to happen because the contractor is more in control of the end costs and if they can control what the design is they will try and make it easier for themselves. And if making it easier means omitting some of the harder aspects [for example, innovative aspects], then that's what will happen.

All architects interviewed expressed the view that 'design and build' procurement inhibits the implementation of sustainable innovation mainly because contractors are all too happy to settle for building adhering to the minimum regulatory standards. Another architect told us:

If you are in a 'design and build' arrangement the first thing [the contractor] does is equate building regulation requirements with the [architect's] innovative specification, which is way above the building regulations. Once he finds out contractually what his requirements are then that's what you build. When 'buildability' brings in an easier way of doing it by omission or by diluting a certain specification, then that's frustrating.

One of the more attractive characteristics of traditional procurement, in terms of implementing sustainable innovations, is the lead role of the architect and consulting engineer. This is important, again, particularly with sustainable technologies, because of the specialized knowledge required – capabilities more likely to be found in an architectural or engineering practice. Their greater involvement, earlier in the process, can be very influential and is possible outside traditional procurement practices but requires integration of the construction team with contractors and subcontractors working closely with the architect and engineer. Some of the representatives of housing associations believed that for this reason, innovative projects should be procured traditionally, arguing for example:

… architects don't like to have a 'design and build' developer because [their design] gets watered down and [contractors] will do what they have to do and no more. And if they can cut corners it increases their profit and that's the [nature of the] business they run. So it definitely works better having traditional procurement on these sustainable developments which kind of bucks the trend on other things.

More recent forms of contract for the consulting engineer and architect have seen them transferred from the employment of the client to the contractor. This has serious implications for innovation if the contractor will not accept responsibility for the innovative systems design of the consulting engineer. The problem arises because the 'fit for purpose' definition is far more stringent than the 'best professional advice' definition, as argued by a consulting engineer interviewed:

Unless somebody, and it has to be the client, is prepared to accept a normal 'professional advice' type arrangement for the innovative systems in the building it is probably an insurmountable obstacle. Any consulting engineer who decided to take it on and ignore it is taking a huge risk. We do 'design and build' work all the time and normally we are quite happy with the 'fitness for purpose' arrangements because we're not doing anything unknown. But, [with an innovative system], any contractor wanting that system to be 'fit for purpose', would be [crazy] because I can't guarantee that it is 'fit for purpose' and I designed it.

Indeed, different definitions are required in the contracts between different parties. For example, a 'fitness for purpose' definition is often required in a contract between an engineer and a housing association, whereas a 'best professional advice' is required in a contract between an engineer and a contractor. This may cause tensions and conflicts in different organizational arrangements, especially with regard to innovative designs.

Implications

Overall, a number of alternative organizational and procurement forms have been used over recent years to promote the adoption of sustainable technologies. However, there seems to be no agreement among the different parties in the construction process about the superiority of any one contractual arrangement. The traditional form of contract is regarded as having advantages in terms of giving the specialist architects and engineers time to devise innovative sustainable solutions. Under 'design and build' type arrangements, earlier involvement of the contractor appears to have advantages in terms of improving the buildability of projects and, arguably, controlling costs and facilitating a faster delivery of the project.

However, while there is a lack of agreement regarding the preferred procurement form, there does seem to be agreement among the different parties that long-term relationships are important for the introduction of sustainable technologies since they foster trust, stability and economies of learning and experience. With closer inter-organizational relations, time and cost improvements are forthcoming. A representative of a housing association explains this in simple terms:

Long-term relationships make projects easier to build, price and design ... creating an environment where you develop trust, [beyond which] you can go on to innovate in other things.

Developing long-term relationships has been a key priority for Scottish Homes to facilitate innovation. Documents describe alternative arrangements, from informal to formal, including 'project partnering' (in one-off projects) or 'strategic partnering' (in multi-phased projects) (Scottish Homes 2000c) – through which closer ties are established with contractors, architects, engineers and other parties in the construction process. The aim of this approach lends itself to sustainable projects since innovations often stem from suppliers and housing associations do not tend to get involved in the supply chain below the main contractor. There are a number of examples of how this approach has been applied in practice. For example, the concept of partnering has been prototyped by a number of housing associations that conducted what they called 'partnering briefing meetings' early in the building process. The aim of these meetings, attended by sub-contractors and suppliers, was, in their words, to 'integrate the team' and explain the philosophy of the project so all actors involved in the construction process were well briefed on the sustainable aims and objectives. Also, relations with the users of the sustainable technologies provides an important feedback mechanism to ensure the continued success in the implementation and diffusion of sustainable technology in construction. However, as we have seen above, there remain conflicts of interest among the different parties that militate against the achievement of policy aims to promote the use of sustainable technology.

PROMOTING INTER-ORGANIZATIONAL COLLABORATION AND SUSTAINABLE TECHNOLOGIES

National housing agencies, such as Scottish Homes, which assists approximately one-third of all house building in Scotland, have a clear ability to influence construction practices to support sustainable building in the social housing sector. However, despite the Agency's acknowledged priority to improve housing quality whilst operating within a limited funding regime and notwithstanding its 'Sustainable Development Policy' (Scottish Homes 2000b) and 'Sustainable Housing Design Guide' (Scottish Homes 2000a), housing associations argue that they do not receive adequate support, either in financial or in risk-sharing terms to implement sustainable technologies. Although national housing agencies have to operate within budgets set by their national governments, they have the autonomy to channel funding to specific areas. So, for example, while Scottish Homes has no plan to increase social housing grants across the board, additional funding could be targeted to particular regions of Scotland or to encourage specific technologies (for example, sustainable technologies with higher capital costs but lower running and maintenance costs). To facilitate this process, national housing agencies and public sector funding in general must begin to consider lifecycle or environmental and social costs.

Innovation competitions play a very important role in prototyping technologies and techniques that would otherwise not be implemented. The use of technologies is stimulated not only within competition projects, but also in subsequent conventionally funded projects. For example, the Scottish Homes 1997 innovation competition not only directly funded several 'winning' projects, but also subsequently funded 'rejected' specifications under the conventional HAG system (the Housing Association Grant, a previous incarnation of the Social Housing Grant). The results from our interviews show that many of the differences in the interests of the different parties in construction would be reconciled if there was more specific funding channelled toward integrating innovative products (for example, through changing procurement criteria to encourage product differentiation and wider technology adoption) and processes (for example, through promoting modernized production methods) and to establishing procedures to assess these innovations. The different organizations interviewed argued that it would help if public funding bodies had clear and different consideration of costs and time in projects using sustainable technologies from those using incumbent technologies.

Equally important in reducing some of the tensions between the parties in the construction process appears to be the repetition of successful demonstration projects. Innovative pilot projects will always be more expensive, particularly where sustainable innovation is concerned, since the market for sustainable technologies is under-developed, supply is constrained and the technology is unfamiliar to all but a few specialists. National housing agencies, in conjunction with their operational partners, need to evaluate funded projects that have used new technology, to determine best practice and disseminate guidance and recommendations. The organizations interviewed stressed the need for simple advice regarding sustainable technologies to be issued to all parties in the construction industry with other publications aimed at the public in general and social housing tenants in particular. Innovative pilot projects require a specification that, on the whole, can feed into the housing association's core specification. Emphasis may be placed on supporting the replication of technologies that could make a difference to many homes and to repeat the specifications. There was agreement among the different organizations interviewed that this may be the best use of limited resources to combat fuel poverty and to achieve sustainable housing. Replication may allow specialist firms to develop and generalist firms to have some experience of sustainable issues. The market for sustainable construction services (design, consultants, contractors and suppliers) may be stimulated, reducing the time and cost barriers.

The sustainability and ecological requirements of social housing developments, (for example, Scottish Homes' 'Sustainable Development Policy' (Scottish Homes 2000b)) represent a new commercial opportunity that may be exploited by construction professionals. Contractors with experience of working with a sustainable supply chain or managing and controlling a sustainable building process may be able to gain a competitive advantage, particularly in 'design and build' contracts. Moreover, consultants that can gain experience working on energy efficiency and sustainable projects in the public sector, may, in the future, when technology prices are lower, be able to apply their knowledge and educated risk assessment in mainstream private sector projects.

Housing associations play a crucial role initiating a sustainable agenda. They not only determine the building specification, but also have control over the commissioning of designers, consultants, contractors and have autonomy over the type of procurement initiative employed. Our interviews with the housing associations demonstrate that alternative types of procurement have been used with mixed success. Project officers have different opinions on the value of increasing the role of the contractor and other alternative procurement arrangements. However, all parties supported the advantages of fostering innovation within long-term inter-organizational collaborative relationships.

A number of measures may help reduce the tensions surrounding the implementation of sustainable technologies. First, it may be useful to include the project coalition at an early stage in the construction process. Second, learning and experience may be diffused more easily by replicating demonstration projects. These practices may reduce some of the tensions between the different aims of the various parties in the construction process and may help to overcome some of the barriers to the achievement of policy aims of sustainability.

CONCLUSIONS

This chapter is concerned with improving our understanding of the fact that the process of innovation often involves the participation of several firms and organizations. Much of the innovation literature sees the innovating firm as the sole locus of the innovation process, and then adds on cooperation for innovation in an ad-hoc way. The general problem with this approach is that it neglects the complex array of mechanisms through which inter-organizational coordination is achieved (and which cannot be reduced to either market transactions or formal contracts). These limitations are demonstrated through the above analysis of evidence from the construction industry, drawing on the particular case of innovation in sustainable technologies.

In the construction industry, the interactions and interdependencies between organizations (including contractors, government, clients, designers, subcontractors, suppliers and tenants) have an important role in shaping the process of production and innovation. The successful adoption of innovations depends, in part, on the efficient and co-operative functioning of the whole network. However, the evidence from the Scottish Homes policy of promoting sustainable technologies reveal factors that enabled and hindered innovation. On the one hand, the Scottish Homes Agency was very proactive in driving policy initiatives to promote sustainable products and processes. As such, all the conditions for the successful introduction and diffusion of innovation appear to be present – an industry characterized by a network form, backed up by funding from a body with a strong strategic focus. On the other hand, a number of factors conspired against innovation – factors that in fact lie at the heart of the particular network form characteristic of the construction industry. Relations between housing associations and contractors were marred by conflicting profit and non-profit motives, which led to differences in the willingness to develop alternative procurement forms. Representatives from the design teams (architects and consulting engineers) added an additional element of conflict due to their favouring traditional procurement as a better condition for innovation. Finally, the importance of relations with tenants means that innovation must also be accompanied by education packs if the technologies are to be applied effectively.

In conclusion, therefore, despite the strong adherence of a major housing agency (Scottish Homes) to a policy of promoting sustainable technologies, in fact it was the very characteristics of the network form of the construction industry that appears to conspire against innovation. Regardless of policy initiatives to promote sustainable technologies and processes, the different aims of the parties involved in the network may not be easily reconciled and traditional approaches to construction may reinforce these differences hindering the effects to introduce innovation through construction networks.

NOTES

1. All parties interviewed were asked to identify the principal sources of sustainable innovations and to identify from their experience what factors facilitated or inhibited the use of sustainable technologies in social housing. In addition, all parties were asked questions about their role in driving innovation (for example, identifying and promoting the use of new technologies), their interaction with other parties in the building chain and their assessment of the role of other parties in driving innovation. They were also asked about the impact of different procurement regimes, and the impact of long-term relationships in terms of encouraging successful sustainable technologies. The interviews, which lasted between one hour and three hours each, were conducted on site in Scotland during the summer of 2000 with managers from the national housing agency and their partner organizations, development officers from housing associations, partners in architectural, engineering and building contractor firms and managing directors of private house builders.

2. This chapter makes reference to three such innovative competition schemes: the scheme at Glenalmond Street, Glasgow (commissioned by Shettleston Housing Association), Love Street, Paisley (commissioned by Williamsborough Housing Association) and Nelson Street, Perth (commissioned by Perthshire Housing Association). These social housing projects

received additional funding either from Scottish Homes or the EU. Further information on these projects and many other examples of innovative building in Scotland in the social housing sector can be found in Scottish Homes (2000a), Ecorenewal (2000) and Gilbert (2000a and 2000b).

3. At the national level, the Climate Change Draft UK Programme (DETR, 2000a) explains the overall strategy to deliver the UK's legally binding target from the Kyoto Protocol to cut greenhouse gas emissions to 12.5 per cent below 1990 levels by 2008–12. One section of the report is concerned with the domestic sector, discussing ways in which to improve energy efficiency in housing whilst delivering benefits to people on low incomes. Proposals include better advice and information, incentives, regulation and partnerships with, for example, local authorities and registered social landlords (housing associations). Also, DETR (2000b) included proposals for fiscal measures (for example, the landfill tax), changes to public sector procurement, development of the construction industry's image, waste minimization and resource conservation. In addition, EESOP (Energy Efficiency Standards of Performance Scheme), run by public electricity suppliers but overseen by the regulator, initiated in 1994 finances measures and equipment to increase household energy efficiency through a £1 levy on all households' annual electricity bills. At the local authority level, HEES (Home Energy Efficiency Scheme) pays for the installation of energy-saving measures in households receiving state benefits. Under HECA (Home Energy Conservation Act), local authorities are required to have a cost-effective strategy to raise the energy efficiency of households in their area. For more information see Royal Commission on Environmental Pollution (2000).

4. For example, the first 'Training Programme on Sustainability', funded jointly by Scottish Homes and the SFHA, ran between February 2001 and May 2001.

5. For new-build, the Scottish Homes subsidy level is 70 per cent with the remaining 30 per cent coming from private loans on the strength of the net rental income from the project.

6. Scottish Homes has also led the housing associations into innovative design through initiatives such as 'Secure by Design' (a crime initiative) and 'Housing for Varying Needs' (a disabled access initiative). Ultimately however, the responsibility rests with the housing association or the project managers since it is they who employ the design and construction team and they who provide a specification document that buildings must meet.

6. Conclusion

Construction firms are embedded in a tight network of organizations which includes other industries in the supply stream, end-users as well as government agencies. In this network, contractors are important sources, and adopters, of innovations that improve construction technologies and act as important mediators of the different flows of technology and information in the construction industry. However, the extent to which contractors and other organizations play a major role in the development and diffusion of innovation, depends upon a number of factors. These include the form of ownership and management structures and the type of networks of relations and interactions with other contractors, professionals, subcontractors, suppliers, the government and research institutes and universities. Our research shows that these factors vary across different European countries but in each case act as important determinants of the nature and extent of development and adoption of innovation in construction.

IMPLICATIONS FOR POLICY

Policies and initiatives to improve the adoption and diffusion of innovation in construction must recognize this finding explicitly. The mere availability of innovative construction products and processes is not a sufficient condition for their adoption and diffusion. Certain forms of ownership and management structures, along with particular types of network relations, facilitate investment in new technologies. In their absence, contractors and other organizations must be assisted by appropriate, and wide-ranging, government initiatives.

The evidence from the five European countries presented in this book shows that government can play an important role in supporting the development and adoption of innovations through a number of measures:

- guaranteeing public markets for innovative firms;
- setting an example to industry by supporting alternative procurement relations;
- acting as a broker to bring together collaborations and networks across construction firms; and

- facilitating relations between contractors and a wide range of institutions such as universities and specialist subcontractors.

Although ownership and financial features of a country are difficult to reshape, government can have an important role in guaranteeing public markets for innovative firms. Governments can set an example to industry by supporting alternative procurement relations. Thus, government can take steps to create a supportive environment in which inter-organizational co-operation can develop more effectively. In this role, it can ensure that the benefits of adopting innovations spill over to the weaker organizations in the network.

Sustainable innovations are at an even greater disadvantage because of the higher costs and uncertainty regarding their reliability compared to incumbent systems. In order to implement sustainable technologies successfully the whole project coalition has to be included. Also, knowledge and experience must be diffused among all parties. Clients, consultants, builders and tenants must be educated. Arguably, and, most importantly, pilot projects must be replicated. Replication will allow specialist firms to develop and generalist firms to have some experience of sustainable issues. The market for sustainable design, consulting services, constructing and supplying will be stimulated across the public–private sector divide. The construction industry offers good examples of increased resource productivity and lower finished product total cost in the presence of stricter environmental regulation. More stringent regulations applied to new housing and improvements to the existing housing stock is needed to stimulate sustainable innovation and create demand for higher-priced alternatives.

Governments can draw lessons from policies and measures already in place in different European countries, such as:

- fixed-price schemes for sustainable technologies;
- direct capital grant support and tax incentives for sustainable energy projects;
- net metering to encourage small-scale renewable energy production;
- low-interest loans for sustainable technologies; and
- lower mortgage rate for buildings that will improve the quality of the built environment (for example, energy efficient buildings and healthy housing).

Governments can address the problems through a battery of measures including regulation, taxation and legislation and can encourage innovation by acting as a broker in a 'technology procurement' strategy. In addition, government may introduce grants and fiscal incentives, channelling funds toward R&D and facilitating economies of learning and experience, beginning with demonstration projects and continued through information dissemination. This dissemination of information has to be aimed at the construction industry, the public in general and social housing tenants in particular.

It is clear that housing associations play a vital role in supporting the innovation process since they have much of the control over the commissioning of designers, consultants and contractors in social housing. Only if they are engaged and convinced of the desirability of energy efficiency will finance be raised and sustainable housing designed and produced. Indeed, one could consider the need for acknowledging sustainability as an indicator for all housing developments, applied to new-build and, retrospectively, to retrofit building work. Housing associations have been encouraged to conduct energy efficient and environmentally sound refurbishment of their existing stock. Also, a more proactive stance on behalf of home owners and tenants is required. This is likely to require substantial subsidies from government and local authorities or industry to projects where particularly innovative technologies or construction methods are used to increase energy efficiency or to reduce the environmental impact. Also, an alternative transparent set of measures for the evaluation of the costs and time in projects using sustainable technologies rather than incumbent technologies needs to be devised. In particular, more emphasis should be placed on life-cycle costs versus capital costs housing associations need to appraise and repeat successful pilot projects that use sustainable technologies, disseminating best practice and promoting integration of successful innovation into housing association specification documents.

When the technologies have become more widely used in social housing projects, the experience embedded in the network of construction firms can be translated across to the private sector. This will apply mainly to new housing. Increased sustainability in the existing housing stock can be encouraged by targeting home owners. This involves a role for central and local government to facilitate information on the advantages of sustainable technologies and to provide fiscal incentives (such as grants and tax breaks) to encourage their diffusion.

IMPLICATIONS FOR THEORY

Empirical research on innovation has neglected issues of corporate strategy and structure. Also, it has paid inadequate attention to the relations between organizations in the process of innovation. The way in which the internal and external organization of firms affects the development of innovation capabilities and the complex array of mechanisms through which inter-organizational co-ordination is achieved has been neglected. The comparative evidence we present on the effect of both corporate governance and networks on innovation in the construction industry in five European countries highlights the variety in firm structure and strategy and the different strategic approaches to networks. Both corporate governance and networks differ (despite similar sector conditions) according to the nature of the different national institutional frameworks within which firms' production activities are conducted.

The peculiar nature of construction – especially its fragmentation, discontinuity and project-based structure – may present a problem for the accumulation of knowledge.

This calls for an examination of three factors that may play an important role in innovation in construction:

- the structure of ownership and management;
- the creation of institutions within the firm to facilitate the diffusion of new processes and practices across the different divisions; and
- long-term relations between firms and collaborations with external sources of knowledge.

Our results confirm that incremental technological change may be supported by certain forms of ownership and management, namely, concentrated ownership and decentralized management. Also, governance systems with cross-holdings among industrial firms, which facilitate long-term relations between them, are also conducive to firm-specific irreversible investments in construction.

The results of this study suggest that studies of innovation need to examine in more detail the following features, which may affect the readiness of firms to invest in innovation:

- ownership and control;
- income derived from abroad;
- management structure; and
- form of cross-holdings.

Our study also suggests that the effective adoption of innovation, and particularly environmental innovation, requires the participation and collaboration of all parties in the industry. Because unlike many other industries, innovations in construction are not implemented within construction firms themselves but on the projects on which firms are involved, the management of innovation in construction is complicated by inter-firm co-ordination and demands along the building chain. The strength of inter-organizational co-operation may be responsible for enhanced performance of the construction industry in some countries. Because construction firms relate to many other industries in the supply stream, together with clients and with government through particular technology and information flows, construction industry innovation can only be understood in relation to the networks in which construction firms are embedded.

Innovation studies needs to focus more deeply on the range of stable relations that allow feedback processes and enable non-routine, flexible behaviour, and risky investments in new product and process technologies. In our study, we concentrated on the following relations:

- relations between contractors and subcontractors or suppliers of materials;
- relations between contractors and the government (for example, in its regulatory role, or in its encouragement of demonstration projects);
- relation between contractors and universities;
- relations between contractors and architects or engineers;
- relation between contractors and clients; and
- international collaborations among contractors.

More internationally comparative research is required to understand the relations between the parties in industries such as construction both on and off site and at different sectoral boundaries that contribute to innovation.

Appendix

Table A.1 Contractors interviewed

Firms	Title(s) of person(s) interviewed and division	European ranking in 1997 (1992)	Turnover £m 1997 (1992)	Profit £m 1997 (1992)	No. of employees 1997 (1992)
Denmark					
Hojgaard & Schultz	Housing Division Director	119 (120)	317 (251)	1.6 (–14)	2828 (3035)
Monberg & Thorsen	Chief Engineer	160 (194)	228 (136)	5.2 (1)	2413 (2036)
Skanska Jensen	1. Head of Renovation 2. Project Engineer	169 (–)	217 (–)	–1.8 (–)	1841 (–)
NCC Danemark	Head of Building Renovation	241 (–)	130 (–)	–0.9 (–)	2041 (–)
France					
Bouygues	Managing Director (Technical) *Bouygues Construction*	1 (1)	9196 (6756)	82 (74)	91979 (83699)
Groupe GTM	R&D Director *GTM Construction – Building*	4 (9)	5459 (2991)	–5.6 (22)	66796 (46993)

France	SGE	1. Technical Director 2. Scientific R&D Director *SGE Campegnon Bernard*	5 (2)	5370 (4786)	−46.6 (44)	54838 (66077)
Germany	Philipp Holzmann	Head of R&D Co-ordination President of ENCORD	2 (3)	6048 (4507)	−0.8 (98)	51200 (43680)
	Hochtief	1. Head of R&D Co-ordination 2. R&D Consultant	6 (11)	5192 (2724)	62.3 (104)	40800 (29254)
	Strabag	Director of Business Development Chairman of European Construction Institute	11 (20)	2669 (1697)	4.2 (31)	19900 (20035)
Sweden	Skanska	Vice President *Skanska Teknik*	7 (8)	5018 3090	494.3 (−332)	32278 (28646)
	NCC	1. Technical Director 2. General Manager Corporate R&D *NCC Teknik*	17 (17)	2197 (1946)	44.2 (−119)	15964 (17561)
	PEAB	Head of Purchasing	39 (−)	1135 (−)	9.9 (−)	7535 (−)
	JM	1. President 2. Head of Building	67 (−)	232 423	64.3 (9)	1930 (2860)

Continued overleaf

Firms	Title(s) of person(s) interviewed and division	European ranking in 1997 (1992)	Turnover £m 1997 (1992)	Profit £m 1997 (1992)	No. of employees 1997 (1992)
AMEC	Director of Technology and Innovation	10 (14)	2774 2122	9.7 (−88)	24269 (27145)
UK John Laing	Head of Design Management	34 (27)	1255 (1270)	16.1 (12)	8393 (9600)
Tarmac (Carillion)	Chief Engineer	26 (10)	1535 (2935)	3.6 (−350)	12348 (28590)

Note: European ranking in terms of turnover.

Source: Building (1993, 1997, 1998), individual firms' Annual Reports (1998).

Table A.2 Contractors interviewed and construction industry statistics 1997

	Total country construction output (Pound sterling bn)	Total country employment (thousands)	Number of firms interviewed	Turnover in firms interviewed as percentage of total construction output	Employment in firms interviewed as percentage of total industry employment
Denmark	12.74	158	4	7.00	5.8
France	60.96	1420	3	32.95	15.0
Germany	172.04	2937	3	8.00	3.8
Sweden	13.84	217	4	62.00	26.6
UK	58.00	1390	3	9.60	3.4

Source: FIEC (1999), individual firms' annual reports (1998).

Table A.3 Other organizations interviewed

	Name of organization	Type of organization	Title of person interviewed
Germany	Bund Deutscher Architekten (BDA)	Architect Trade Association	1. Legal Advisor 2. Member of BDA Presidential Board
	Zentralverband Deutches Baugewerbe (ZDB)	Trade Association	1. Head of the Economics Division 2. Technical Engineer
Sweden	Ahlqvist & Co – Arkitecter	Architect	Partner/owner
	Arkitekt & Ingenjorsforetagen	Swedish Federation of Architects and Consulting Engineers	1. Chief Engineer 2. Chief Architect
	Byggforskningsradet (BFS)	Swedish Council for Building Research	1. Director General 2. Head of Department
	Byyggstandardiseringen (BST)	Swedish Building Standards Institute	Project Manager
	Byggkostnadsdelegationen	Government Commission on Costs in Construction	Industry Minister
	SBUF	Development Fund of the Swedish Construction Industry	President

Denmark	C.F. Mollers Tegnestue	Architect	Senior Partner
	Danske Entreprenorer	Trade Association	Economist
	Danish Ministry of Housing	Government	Head of Development Division
	SBS Byfornyelse	Non-Profit Housing Association	Managing director
	Statens Byggeforskningsinstitut (SBI)	Danish Building Research Institute	1. Civil Engineer 2. Chief Economist Building Productivity
UK	Gifford & Partners	Engineer	Partner
	WS Atkins	Engineer	Director of Innovation

Table A.4   Organizations interviewed involved in Scottish social housing

Organization name	Organization type
Ballast Wiltshire plc (Scotland) Melville Dundas Morrison Construction Group plc	Contractor
Beazer Homes Bett Partnership Persimmon Homes Partnership Housing	Private Sector Developer
Canmore Housing Association Castle Rock Housing Association East Lothian Housing Association Edinvar Housing Association Kingdom Housing Association Meadowside and Thornwood Housing Association Perthshire Housing Association Shettleston Housing Association Thenew Housing Association Williamsborough Housing Association	Housing Association
Carl Bro Enconsult Waterman Gore	Consulting Engineer
James F. Stephen Architects John Gilbert Architects Maclachlan and Monahon Architects Murray Design Group Riach Partnership	Architect
Scottish Federation of Housing Associations	Housing Association's Trade Association
Scottish Homes	National Housing Agency

References

Abernathy, W. and J.M. Utterback (1978), 'Patterns of industrial innovation', *Technological Review*, **80** (7), 41–7.

Adolfsson, P., U. Holmburg and S. Jonsson (1999), 'Corporate governance in Sweden – a literature review', submission to the EU Commission as part of the Corporate Governance and Product Innovation Project, accessed 24 July, 2003, at www.sums.ac.uk/copi/reports/corpgov/sweden.htm.

Allder, G. (1999), 'Fill the cavity', *Housing Association Building and Maintenance*, **8** (2), 15–16.

Allinson, K. (1993), *The Wild Card Design: A Perspective on Architecture in a Project Management Environment*, Oxford: Butterworth Architecture.

Arditi, D., S. Kale and M. Tangkar (1997), 'Innovation in construction equipment and its flow into the construction industry', *Journal of Construction Engineering and Management*, **123** (4), 371–8.

Arthur, W.B. (1988), 'Competing technologies: an overview', in G. Dosi, C. Freeman, R. Nelson, G. Silverberg and L. Soete (eds), *Technical Change and Economic Theory*, London: Pinter.

Atkinson, A.B. and J.E. Stiglitz (1969), 'A new view of technological change', *Economic Journal*, **79** (315), 573–8.

ATV (1999), *Byggeriet I det 21*, Lyngby, Sweden: Arhundrede.

Ball, M. (1988), *Rebuilding Construction: Economic Change in the British Construction Industry*, London: Routledge.

Ball, M. (1996), *Housing and Construction – A Troubled Relationship*, Bristol: Policy Press.

Bell College of Technology (1994), *Petra Euromodule 1994 Life Cycle Assessment Example – Thermal Insulation*, Hamilton, Scotland: Bell College of Technology.

Bertlesen, S. and J. Nielsen (1999), 'The Danish experience from 10 years of productivity development', paper presented at the 2nd International Conference on Construction Industry Development, October.

Best, M. (1990), *The New Competition: Institutions of Industrial Restructuring*, Cambridge, MA: Harvard University Press.

Blackley, D.M. and E.M. Shepard (1995), 'The diffusion of innovation in home building', *Journal of Housing Economics*, **5**, 303–22.

BMBF (1999), Leitprojekte, accessed 24th July, 2003, at www.bmbf.de.

Bokhoven, T. (2001), 'Solar thermal market development', *Renewable Energy World*, July–August, 231–9.

Bordass, B. (2000), 'Cost and value: fact and fiction', *Building Research and Information*, **28** (5/6), 338–52.

Bosselaar, L. (2001), 'Solar heating – a major source of renewable energy', *Renewable Energy World*, July–August, 219–29.

BRE (Building Research Establishment) (1999), 'Sustainable construction', BRE client report no. CR 258/99, accessed 24 July 2003, at www.bre.co.uk.

Breschi, S., F. Malerba and L. Orsenigo (2000), 'Technological regimes and Schumpeterian patterns of innovation', *Economic Journal*, **111** (463), 388–410.

Buckley, P.J. and M. Casson (1990), 'Joint ventures', in M. Casson (ed.), *Enterprise and Competitiveness*, Oxford: Oxford University Press.

Building (1988), 'Top 300 European contractors', Procurement Supplement, *Building*, December, 37–41.

Building (1994), 'Eurofile Germany', *Building Economist*, supplement, *Building*, September, 10–11.

Building (1997), 'Top 300 European contractors', Procurement Supplement, *Building*, December, 52–9.

Building (2000), 'War on waste', *Building*, 18 July, 18–21.

Cadman, D. (1999), 'Environmental audits of the construction industry: what they show and how they can be broadened and acted upon', proceedings of the Construction Confederation Conference: Constructing a Sustainable Environment, Birmingham.

Caleb Management Services (1997), *Thermal insulation and its Role in Carbon Dioxide Emission Reduction*, Bristol: Caleb Management Services.

Carillion (2000), 'We are making choices: Carillion's environment, community and social report 1999–2000', accessed 24 July 2003, at www.carillionplc.com.

Castells, M. (1996), *The Rise of the Network Society*, Oxford: Blackwell.

Cleff, T. and A.R. Cleff (1999), 'Innovation and innovation policy in the German construction sector', working paper, Centre for European Economic Research (ZEW).

Coase, R. (1937), 'The nature of the firm', *Economica*, 4, 386–405.

Construction Resources (2000), 'Construction Resources product data: natural insulation', accessed 24 July 2003, at www.constructionresources.com.

Coombs, R., M. Harvey and B. Tether (forthcoming), 'Analysing distributed processes of provision and innovation', *Industrial and Corporate Change*.

Cowe, R. and S. Williams (2000), 'Who are the ethical consumers?', accessed 5 November 2003, at www.co-operativebank.co.uk.

CPB (Netherlands Bureau for Economic Policy Analysis) (1997), *Challenging Neighbours: Rethinking German and Dutch Institutions*, Berlin: Springer-Verlag.

Curwell, S.R. and C.G. Mach (1986), *Hazardous Building Materials*, London: E&FN Spon.

Curwell, S.R., C.G. Mach and D. Venables (1990), *Buildings and Health – the Rosehaugh Guide to the Design, Construction, Use and Management of Buildings*, London: RIBA Publications.

David, P. (1985) 'Clio and the economics of QWERTY', *American Economic Review*, **75** (2), 332–37.

Davies, A. and T. Brady (2000), 'Organisational capabilities and learning in complex product systems: towards repeatable solutions', *Research Policy*, **29** (7–8), 931–53.

Deakin, S. and F. Wilkinson (1998), 'Contract law and the economics of inter-organizational trust', in C. Lane and R. Bachmann (eds), *Trust Within and Between Organizations*, Oxford: Oxford University Press.

DETR (Department of Environment, Transport and the Regions) (2000a), *UK Climate Change Programme*, London: DETR.

DETR (Department of Environment, Transport and the Regions) (2000b), *Building a Better Quality of Life: A Strategy for more Sustainable Construction*, London: DETR.

DETR (Department of Environment, Transport and the Regions) (2000c), *The Building Act 1984 Building Regulations: Proposals for Amending the Energy Efficiency Provisions*, London: DETR.

Donaldson, L. and J.H. Davis (1994), 'Boards and company performance – research challenges the conventional wisdom', *Corporate Governance*, **2** (3), 151–60.

Eccles, R. (1981), 'The quasi-firm in the construction industry', *Journal of Economic Behaviour and Organisation*, **2**, 335–57.

Ecorenewal (2000), 'Ecological building projects in Europe', accessed 5 November 2003, at www.ecorenewal.com.

Edwards, D. (1998), *The Link Between Company Environmental and Financial Performance*, London: Earthscan Publications.

Edquist, D. (1997), *Systems of Innovation: Technologies, Institutions and Organizations*, London: Pinter.

Egan, J. (1998), *Rethinking Construction: Report of the Construction Task Force*, London: HMSO.

Environmental Data Service (ENDS) (1998), 'CO_2 target to drive change in buildings energy efficiency', ENDS report 282.

ENERDATA/Odyssee (1999), 'Energy efficiency in the European Union 1990–1998', accessed 24 July 2003, at www.odyssee-indicators.org/Publication/ PDF/Mon_overall.pdf.

EREC (1995a), 'Consumer energy information: loose fill insulation, energy efficiency and renewable energy networks', US Department of Energy, accessed 3 November 2003, at www.eere.energy.gov/consumerinfo/factsheets/insulate.html.

EREC (1995b), 'Consumer energy information: EREC reference briefs – radiant barriers, energy efficiency and renewable energy networks', US Department of Energy, accessed 3 November 2003, at www.eere.energy.gov/consumerinfo/rebriefs/bc7.html.

ETSU (1999a), 'Large scale solar purchasing', ETSU: ETSU S/P3/00266/REP, available on loan from ETSU, www.etsu.com.

ETSU (1999b), 'Active solar heating system performance and data review', ETSU: ETSU S/P3/00270/REP, available on loan from ETSU, www.etsu.com.

EurObserv'ER (2000), 'EurObserv'ER Barometer', *Renewable Energy Journal*, **10**, 32–41.

EEA (European Environmental Agency) (2001a), 'Europe's environment – the Dobris assessment, households, Chapter 26', accessed at www.themes.eea.eu.int/sectors_and_activities/household/reports.

EEA (European Environmental Agency) (2001b), 'Europe's environment – indicators', accessed at www.themes.eea.eu.int/sectors_and_activities/household/indicators.

European International Contractors (EIC) (1998), *Mergers and Acquisitions of the European Construction Industry: Holdings in Europe and Overseas*, 7th edn, Berlin: EIC.

European Solar Industry Foundation (1995), 'Sun in action – the solar thermal market: a strategic plan for action in Europe', ATENER report 4.1030/E/94-003.

Eurostat (1998), *Labour Force Survey*, Luxembourg: Eurostat.

Everett, B. (1996), 'Solar thermal energy', in G. Boyle (ed.), *Renewable Energy: Power for a Sustainable Future*, Oxford: Alden Press.

Fama, E. and M. Jensen (1983), 'Separation of ownership and control', *Journal of Law and Economics*, 26, 301–25.

European Construction Industry Federation (FIEC) (1999), *Construction Activity in Europe*, Brussels: FIEC.

Freeman, C. (1987), *Technology Policy and Economic Performance: Lessons from Japan*, London: Pinter.

Freeman, C. (1996), 'The greening of technology and models of innovation', *Technological Forecasting and Social Change*, **53**, 27–39.

Gann, D.M. (1994), 'Innovation in the construction sector', in M. Dodgson, and R. Rothwell, (eds), *The Handbook of Innovation*, Aldershot, UK and Brookfield, US: Edward Elgar.

Gann, D.M. (1997), 'Should governments fund construction research?', *Building Research and Information*, **25** (5), 257–67.

Gann, D.M. (1999), 'Can regulations promote construction innovation?', CRISP Commission, accessed 3 November 2003, at www.crisp-uk.org.uk/REPORTS/9912_fr.pdf.

Gann, D.M. (2000), *Building Innovation: Complex Constructs in a Changing World*, London: Thomas Telford.

Gann, D.M. and A.J. Salter (2000), 'Innovation in project-based, service-enhanced firms: the construction of complex products and systems', *Research Policy*, **29**, 955–72.

Gardiner and Theobold (1998), 'International construction cost survey', accessed 3 November 2003, at www.gardiner.com.

Gilbert, J.D. (2000a), 'Geothermal and solar energy initiatives', in C. Buckle (ed.), *Proceedings of Conference C75 of the Solar Energy Society: Renewable Energy for Housing*, Oxford: The Solar Energy Society, pp. 59–64.

Gilbert, J.D. (2000b), 'Scotland the brave: innovation in housing', accessed 5 November 2003, at www.johngilbert.co.uk.

Gray, P. (1989), 'The paradox of technological development', in J.H. Ausubel and H.E. Sladovich (eds), *Technology and the Environment*, Washington, DC: National Academy Press.

Green, K., A. McMeekin and A. Irwin (1994), 'Technological trajectories and R&D for environmental innovation in UK firms', *Futures*, **26** (10), 1047–59.

Green, K., S. Shackley, P. Dewick and M. Miozzo (2002), 'Long-wave theories of technological change and the global environment', *Global Environmental Change*, **12** (2), 79–81.

Groak, S. (1994), 'Is construction an industry? Notes towards a greater analytic emphasis on external linkages', *Construction Management and Economics*, **12**, 287–93.

Groenwegen, J. (1994), 'About double organized markets: issues of competition and cooperation. The Dutch construction cartel: an illustration', *Journal of Economic Issues*, **28** (3), 901–8.

Harland, E. (1993), *Eco-Renovation: The Ecological Home Improvement Guide*, Dartington, Devon: Green Books.

Harper, D. (2000), 'Emission impossible?', *Housing Association Building and Maintenance*, **8** (5), 32–3.

Harvey, R.C. and A. Ashworth (1997), *The Construction Industry of Great Britain*, Oxford: Butterworth-Heinemann.

Heath, P. (1999), 'Energy in the balance', *Local Authority Building and Maintenance*, **15** (2), 13–14.

Hobday, M. (1998), 'Product, complexity, innovation and industrial organization', *Research Policy*, **26**, 871–93.

Hobday, M. (2000), 'The project-based organisation: an ideal form for managing complex products and systems', *Research Policy*, **29** (7–8), 895–911.

Huru, H. (1992), 'The UK construction industry: a continental view', London: Construction Industry Research and Information Association (CIRIA) special publication 82.

Huskinson, D. (1998), 'The role of local authorities in energy conservation – opportunities for solar water heating', in *Proceedings of Conference C72 of the Solar Energy Society: Solar Water Heating: A Hands on Approach*, Oxford: The Solar Energy Society, pp. 27–30.

Institute of Building Control (1996), *Review of European Building Regulations and Technical Provisions: Denmark*, Epsom, Surrey: Institute of Building Control.

Institute of Building Control (1996), *Review of European Building Regulations and Technical Provisions: the Netherlands*, Epsom, Surrey: Institute of Building Control.

Institute of Building Control (1996), *Review of European Building Regulations and Technical Provisions: Sweden*, Epsom, Surrey: Institute of Building Control.

Institute of Building Control (1997), *Review of European Building Regulations and Technical Provisions: UK*, Epsom, Surrey: Institute of Building Control.

Institute of Building Control (1998), *Review of European Building Regulations and Technical Provisions: France*, Epsom, Surrey: Institute of Building Control.

Institute of Building Control (1998), *Review of European Building Regulations and Technical Provisions: Germany*, Epsom, Surrey: Institute of Building Control.

IPCC (Intergovernmental Panel on Climate Change) (1996), 'Economic and social dimensions of climate change', *Climate Change*, vol. 3, IPCC second assessment report, Cambridge: Cambridge University Press.

IPCC (Intergovernmental Panel on Climate Change) (2000), *Emissions Scenarios 2000, Special Report on the Intergovernmental Panel on Climate Change*, Cambridge: Cambridge University Press.

Jaffe, A.B., S.R. Peterson, P.R. Portney and R.N. Stavins (1995), 'Environmental regulation and the competitiveness of US manufacturing: what does the evidence tell us?', *Journal of Economic Literature*, **33** (1), 132–63.

Jensen, M. and W. Meckling (1976), 'Theory of the firm: managerial behaviour, agency costs and ownership structure', *Journal of Financial Economics*, **3**, 305–60.

Kemp, R. (1994), 'Technology and the transition to environmental sustainability', *Futures*, **26** (10), 1023–46.

Kemp, R. and L. Soete (1992), 'The greening of technological progress', *Futures*, June, 437–57.

Kerr, A. and S. Allen (2001), 'Climate change: North Atlantic comparions: the Scottish Executive Central Research Unit', accessed 24 July 2003, at www.scotland.gov.uk/library3/environment/ccna-00.asp.

Kester, W.C. (1992), 'Industrial groups as systems of contractual governance', *Oxford Review of Economic Policy*, **8** (3), 24–44.

Laborde, M. and V. Sanvido (1994), 'Introducing process technologies into construction companies', *Journal of Construction Engineering and Management*, **120** (3), 488–509.

Latham, M. (1994), *Constructing the Team: Joint Review of Procurement and Contractual Arrangements in the United Kingdom Construction Industry*, London: HMSO.

Lazonick, W. (1993), 'Industry clusters versus global webs: organizational capabilities in the American economy', *Industrial and Corporate Change*, **2** (1), 1–24.

Lazonick, W. and M. O'Sullivan (1996) 'Organisation, finance and international competition', *Industrial and Corporate Change*, **5** (1), 1–36.

Lundvall, B.A. (1985), *Product Innovation and User–Producer Interaction*, Aalborg, Denmark: Aalborg University Press.

Lundvall, B.A. (1988), 'Innovation as an interactive process: from user–producer interaction to the national system of innovation, in G. Dosi, C. Freeman, R. Nelson, G. Silverberg and L. Soete (eds), *Technical Change and Economic Theory*, London: Pinter.

Lundvall, B.A. (1992), *National Systems of Innovation: Towards a Theory of Innovation and Interactive Learning*, London: Pinter.

MacGillivray, A. (2000), 'The fair share: the growing market share of green and ethical products', accessed 24 July 2003, at www.co-operativebank.co.uk.

MacGregor, K. (2000), 'Solar energy in Scotland', in C. Buckle (ed.), *Proceedings of Conference C75 of the Solar Energy Society: Renewable Energy for Housing*, Oxford: The Solar Energy Society, pp. 51–2.

Malerba, F. (2002), 'Sectoral systems of innovation and production', *Research Policy*, **31** (2), 247–64.

Malerba, F. and L. Orsenigo (1996a), 'The dynamics and evolution of industries', *Industrial and Corporate Change*, **5** (1), 51–87.

Malerba F. and L. Orsenigo (1996b), 'Schumpeterian patterns of innovation are technology specific', *Research Policy*, **25** (3), 451–78.

Malin, N. (2000), 'The cost of green material', *Building Research and Information*, **28** (5/6), 408–12.

Mallin, C. (1999), 'Financial institutions and their relations with corporate boards', *Corporate Governance*, **7** (3), 248–55.

Miles, R. and C. Snow (1986), 'Organisations, new concepts for new form', *California Management Review*, **28** (3).

Ministry of Business and Industry and Ministry of Housing and Building, (1995), *Process and Product Development in the Building Industry – A Danish Development Programme*, Copenhagen: Danish Agency for Development of Trade and Industry and the National Housing and Building Agency.

Miozzo, M. and C. Ivory (2000), 'Restructuring in the British construction industry: implications of recent changes in project management and technology', *Technology Analysis and Strategic Management*, **12** (4), 513–31.

Nam, C.H. and C.B. Tatum (1988), 'Major characteristics of constructed products and resulting limitations of construction technology', *Construction Management and Economics*, **6**, 133–48.

Nelson, R. (1993), *National Innovation Systems: A Comparative Analysis*, Oxford: Oxford University Press.

Nelson, R. and S. Winter (1977), 'In search of a useful theory of innovation', *Research Policy*, **6** (1), 36–76.

Nelson, R.R. and S.G. Winter (1982), *An Evolutionary Theory of Economic Change*, Cambridge, MA: Harvard University Press.

Nohria, N. and R.G. Eccles (eds) (1992), *Networks and Organizations*, Boston, MA: Harvard Business School Press.

Nyman, S. and A. Silberston (1978), 'The ownership and control of industry', *Oxford Economic Papers*, **30** (1), 74–101.

The Observer (2001), 'PFI's bounty hunters', 8 July, p. 5.

OECD (1995), *National Systems for Financing Innovation*, Paris: OECD.

OECD (1998), *Towards more Sustainable Consumption: Revised Workplan*, Paris: OECD.

OECD (2000), *Analytical Business Enterprise R&D (ANBERD) Database*, Paris: OECD.

Oscar Faber (1998), 'A review of the energy efficiency requirements in building regulations – interim paper', report prepared for the Department of the Environment, Transport and the Regions, London: HMSO.

O'Sullivan, M. (2000a), 'The innovative enterprise and corporate governance', *Cambridge Journal of Economics*, **24**, 393–416.

O'Sullivan, M. (2000b), *Contests for Corporate Control: Corporate Governance and Economic Performance in the United States and Germany*, Oxford: Oxford University Press.

Ouchi, W.G. (1980), 'Markets, bureaucracies and clans', *Administrative Science Quarterly*, **25** (1), 133–48.

Pavitt, K. (1984), 'Sectoral patterns of technical change: towards a taxonomy and a theory', *Research Policy*, **13** (6), 343–73.

Pfeffer, J. and P. Nowak (1976), 'Joint ventures and interorganizational interdependence', *Administrative Science Quarterly*, **21** (3), 398–418.

Piore, M. and C. Sabel (1984), *The Second Industrial Divide*, New York: Basic Books.

Pollington, C. (1999), 'Legal and procurement for sustainable development', *Building Research and Information*, **27** (6), 410–12.

Porter, M. (1990), 'The competitive advantage of nations', *Harvard Business Review*, **90** (2), 73–93.

Porter, M. and C. van der Linde (1995), 'Green and competitive: ending the stalemate', *Harvard Business Review*, September–October, 120–34.

Powell, W.W. (1990), 'Neither market nor hierarchy: network forms of organisation', *Research in Organizational Behaviour*, **12**, 295–336.

Prencipe, A. (2000), 'Breadth and depth of technological capabilities in CoPS: the case of the aircraft engine control system', *Research Policy*, **29** (7–8), 895–911.

Pries, F. and F. Janszen (1995), 'Innovation in the construction industry: the dominant role of the environment', *Construction Management and Economics*, **13** (1), 43–51.

Quigley, J.M. (1982), 'Residential construction', in Nelson, R. (ed.), *Government and Technical Progress: A Cross-Industry Analysis*, New York: Pergamon Press.

RES (1997), 'The European Commission White Paper on Renewable Energies', COM (97) 599, 26/11/97, accessed 24 July 2003, at www.agores.org/policy/com_strategy/white_paper/.

Radice, H. (1971), 'Control type, profitability and growth in large firms: an empirical study', *Economic Journal*, **81**, pp. 547–62.

Rosenfeld, Y. (1994), 'Innovative construction methods', *Construction Management and Economics*, **12** (6), 521–41.

Rothwell, R. (1977), 'The characteristics of successful innovators and technically progressive firms', *R&D Management*, **7** (3), 191–206.

Royal Commission on Environmental Pollution (2000), 'Energy: the changing climate', 22nd report, accessed 24 July, 2003 at www.rcep.org.uk/energy.html.

Scottish Executive (2000), 'Proposed amendments to the Building Standards (Scotland) Regulations 1990: Part J (Conservation of Fuel and Power)', accessed 24 July 2003, at www.scotland.gov.uk/views/consult.asp.

Scottish Federation of Housing Associations (SFHA) (1999), *Raising Standards in Housing*, 3rd edn, Edinburgh: Scottish Homes and SFHA.

Scottish Homes (2000a), *Sustainable Housing Design Guide: A Handbook for Practitioners*, Edinburgh: Stationery Office.

Scottish Homes (2000b), 'Sustainable development policy', accessed 5 November 2003, at www.communitiesscotland.gov.uk.

Scottish Homes (2000c), 'Procurement and partnering: policy advice note', accessed 5 November 2003, at www.communitiesscotland.gov.uk.

Shleifer, A. and Vishny, R. (1997), 'A survey of corporate governance', *Journal of Finance*, **52** (2), 737–83.

Shrivastava, P. (1995), 'Environmental technologies and competitive advantage', *Strategic Management Journal*, **16**, 183–200.

Slaughter, E.S. (1993), 'Builders as sources of construction innovation', *Journal of Construction Engineering and Management*, **119** (3), 532–49.

Slaughter, E.S. (1998), 'Models of construction innovation', *Journal of Construction Engineering and Management*, **124** (3), 226–31.

Slaughter, E.S. (2000), 'Implementation of construction innovations', *Building Research and Information*, **28** (1), 2–17.

Sorrel, S. (2001), 'Making the link: climate policy and the reform of the UK construction industry', SPRU working paper no. 67, July.

Steer, P. and J. Cable (1978), 'Internal organization and profit: an empirical analysis of large UK companies', *Journal of Industrial Economics*, **27**, 13–30.

Sterling, C.M. and B.R. Anderson (1999), 'Review of Part L of the building regulations: technical implications of increased thermal insulation', prepared for Oscar Faber, accessed 24 July 2003, at www.safety.odpm.gov.uk/bregs/ consult/eep/pdf/br05g33.pdf.

Systèmes Solaires (1999), 'Fin d'eclipse', *Systèmes Solaries*, **133**, 1–10.

Systèmes Solaires (2000), 'Thermal solar barometer', *Systèmes Solaires*, **138**, 85–91.

Tatum, C.B. (1986), 'Potential mechanisms for construction innovation', *Journal of Construction Engineering and Management*, **112** (2), 178–87.

Tatum, C.B. (1987), 'Process of innovation in the construction firm', *Journal of Construction Engineering and Management*, **113** (4), 648–63.

Teece, D.J. (1986), 'Profiting from technological innovation: implications for integration, collaboration, licensing and public policy', *Research Policy*, **15**, 285–305.

Thermal Insulation Manufacturers and Suppliers Association (TIMSA) (2000), 'Insulation industry handbook 1999/2000', accessed 3 November 2003, at www.timsa.org.uk/timsa/publications.html.

Thorp, J.P. (2000), 'Max modification: sustainable energy and society', in C. Buckle (ed.), *Proceedings of Conference C75 of the Solar Energy Society: Renewable Energy for Housing*, Oxford: The Solar Energy Society, 39–45.

Toole, T.M. (1998), 'Uncertainty and home builders' adoption of technological innovation', *Journal of Construction Engineering and Management*, **124** (4), 323–32.

Tylecote, A. and E. Conesa (1999), 'Corporate governance, innovation systems and industrial performance', *Industry and Innovation*, **6** (1), 25–50.

UNFCCC (United Nations Framework Convention on Climate Change) (1999), 'Report on the in-depth review of the Second National Communication of Denmark', UNFCCC.

Van der Leun, K. (2001), 'Soltherm Europe Initiative: joining forces to expand solar markets, fast', *Renewable Energy World*, September–October, 127–37.

Van Zee, E. (1999), 'Solar thermal systems by design', in O. Lewis, and J. Goulding (eds), *European Directory of Sustainable and Energy Efficient Buildings* London: James and James.

Von Hippel, E. (1988), *The Sources of Innovation*, Oxford: Oxford University Press.

Weaver, P., L. Jansen, G. van Grootveld, E. van Spiegel and P. Vergragt (2000), *Sustainable Technology Development*, Sheffield: Greenleaf Publishing.

Weimer, J. and J.C. Pape (1999), 'A taxonomy of systems of corporate governance', *Corporate Governance*, **7** (2), 152–65.

Weitzmann, M.L. (1997), 'Sustainability and technical progress', *Scandinavian Journal of Economics*, **99** (1), 1–13.

Welford, R. and R. Starkey (1996), *The Earthscan Reader in Business and the Environment*, London: Earthscan.

White, D. (2000), 'City slackers', *Building*, 10 March, pp. 22–23.

Winch, G. (1996), 'Contracting systems in the European construction industry – a sectoral approach to the dynamics of business systems', in R. Whitley, and P.H. Kristensen (eds), *The Changing European Firm: Limits to Convergence*, London: Routledge.

Winch, G. (1998), 'Zephyrs of creative destruction: understanding the management of innovation in construction', *Building Research and Information*, **26** (4), 268–79.

Winch, G. and E. Campagnac (1995), 'The organization of building projects: an Anglo-French comparison', *Construction Management and Economics*, **13**, 3–14.

Woolley, T., S. Kimminis, P. Harrison and R. Harrison (1997), *Green Building Handbook*, London: E&FN Spon.

Wubben, E. (1999), 'What's in it for us? the impact of environmental legislation on competitiveness', *Business Strategy and the Environment*, **8** (2), 95–107.

Index